SAFETY CASES AND SAFETY REPORTS

To my children
If you can't be safe, at least be careful.
Dad

Safety Cases and Safety Reports
Meaning, Motivation and Management

RICHARD MAGUIRE
B.Eng MSc. C.Eng MIMechE MSaRS

CRC Press
Taylor & Francis Group
Boca Raton London New York

CRC Press is an imprint of the
Taylor & Francis Group, an **informa** business

CRC Press
Taylor & Francis Group
6000 Broken Sound Parkway NW, Suite 300
Boca Raton, FL 33487-2742

First issued in paperback 2017

Version Date: 20160226

ISBN 13: 978-0-7546-4649-5 (hbk)
ISBN 13: 978-1-138-07532-0 (pbk)

Visit the Taylor & Francis Web site at
http://www.taylorandfrancis.com

and the CRC Press Web site at
http://www.crcpress.com

Contents

List of Figures

List of Tables

Preface

The core of safety engineering is a systematic approach to identifying the hazards and hazardous events that could happen, and then eliminating or controlling the risk. All this must be done until the risk is tolerable, and then it must be recorded, demonstrated and sustained over time. In step with this, there is a duty to assure yourself and demonstrate to others that your system, project, process or piece of equipment is tolerably safe, not only to the people who come into direct contact with it, but also members of the public and the environment at large. Corporate image and survivability is at stake, when considering what safety related actions to take. It really can be the difference between life and death, if it is applied correctly, the benefits to the organisation are truly amazing – lower lost time, fewer workplace incidents, improved staff loyalty and a better bottom line.

The major tools for accomplishing all of this is the concept of a safety case and safety report. The parallels with a legal case are useful – they equally scare corporate managers. The main difference with the safety case is that you have the opportunity to construct the case in your own time and whilst you are developing the system. You do not have to be called to court to have to start to prepare your case, you can do it now while you have all the information around you and full control over it.

The key elements of this text are based around identifying the meaning and measurement of safety and risk; the motivation behind the need to construct a safety case; the management of the task of generating and presenting one; and how to maintain it once it has been produced. Explicit guidance is given on developing risk matrices, safety targets, demonstrating ALARP, the value of preventing a fatality and tools and techniques for safety assessments. Coupled with these, are specific chapters on human factors, software factors and management factors and how they influence safety performance and safety cultures. All these areas need to be considered in a robust, consistent and complete safety case.

The text takes a world view of safety engineering across all the hazardous industries – nuclear, rail, chemical, defence and construction, citing historical and not-so-historical incidents to provide real examples of the textural points being made. Some you will probably be familiar with, others show classic traits of poor safety practice and worse safety management.

The importance of the safety case cannot be understated, it has become integral to UK industry, with statutes mandating its use in certain high risk industry. Knowledge of it is required to operate at any level in any of the industries noted above. Additionally, risk and safety have become political issues, the UK Government has expressly said that safety management – getting the right balance between innovation and change on one hand, and avoidance of shocks and crisis on the other – is now central to the business of good government.[1] Within the US, the safety case concept has yet to take real hold. It is certainly known about, even as long ago as 1998, an influential paper on maintaining US leadership in aeronautics directly recommended safety arguments systematically presented in the form of a safety case. This document also cited that this would provide the aircraft industry with an approach to certification that is rapid, repeatable and accurate.[2]

Around the rest of the world, the safety case concepts are being employed with great effect. From European air-traffic control to Australian petroleum facilities, the safety case is in essential use recording risks, the controls in place and the safety management system in place to ensure that the controls are competently and steadily applied.

This book will provide an introduction to and discussion on the contemporary techniques for developing and assessing safety cases and safety reports. It gives an understanding of the principles behind the techniques so that readers can start to make judgements about safety and risk during their studies and work. The text also seeks to enhance the reader's appreciation of the importance of the role of safety engineering within the team, the organisation and the societal community.

Finally, whilst this book offers a full and wide ranging consideration of system safety engineering, it is guidance and discussion only, and is in no way a replacement for full safety assurance. Safety concerns should be addressed by a team of competent professionals, using their experience and judgement in combination with best practice, techniques and other applicable processes.

Richard Maguire
September 2006

---

[1] "Risk: Improving government's capability to handle risk and uncertainty", The Cabinet Strategy Office, 2002.
[2] "Maintaining US Leadership in Aeronautics: Breakthrough Technologies to Meet Future Air and Space Transportation Needs and Goals", The National Academy Press, 1998.

Acknowledgements

The author would like to pass special thanks to all those who have contributed to the content of this book. Special thanks are given to:

The Directors and employees of SE Validation Limited
Members of the Safety-Critical Systems Club
Members of the Safety and Reliability Society.

Also thanks for proofing and having to read the text over and over:

Kirsty Maguire
Colin Brain

Chapter One

Accidents and Safety

Introduction

At whatever stage in your life you are starting to read this book, you will have been aware of disasters in the world. Ever since William Huskisson MP became the first person to be killed on UK Railways in September 1830 on the opening day of the Liverpool to Manchester line, the record of industrial accidents and disasters has been added to with frightening regularity. Even in recent history when disasters have become global media events the list keeps on growing. Table 1.1 contains a list of relatively recent events that may be classed as disasters – certainly by those effected.

Probably everyone reading this now will be recalling memories of these or some dreadful accident that occurred to them, someone they knew, at some place they knew or something else that became a national tragedy, to the extent that it was lead story for days and actually has anniversary memorials. I can think of far too many of these.

However, with each occurrence of harm, injury or loss that takes place, engineers grow more informed about what happens in the world that they build. Design and operating improvements are mandated, codes of better practice are developed and protection and information schemes are put in place. The goal of all these approaches is to not only ensure that similar events do not happen again, but that as time progresses, the world becomes collectively more safe. Each replacement product, system or process should be safer than the one it replaces; each brand new product, system or process should be compared with existing items to benchmark and improve on its safety performance.

Of course it is far better not to have to wait for an accident to occur in order to prevent any similar future ones happening. Humanity is thinking very hard about how accidents initiate, develop and propagate into disasters, such that they can be prevented before they have opportunity to cause harm, injury or loss. Many industries and countries have authorities and inspector organisations that research and police hazardous areas of work and judge safety performance. Evidence is often called for in demonstration of safety performance and this has many beneficial features from identifying areas for improvement to actually providing defence evidence in legal cases.

Table 1.1 Examples of Recent Accidents and Disasters

Industry	Description	Date	Cause(s)	Impact
Rail	Kings Cross	1987	Fire / smoke	31 fatalities
	Arizona	1997	Bridge failure	116 injuries
	Paddington	1999	Training / signal design	31 fatalities, £2m fine
Chemical	Flixborough	1974	Explosion	28 fatalities
	Bhopal	1984	Toxic gas	2500+ fatalities
	Piper Alpha	1988	Fire	167 fatalities
Nuclear	TMI	1979	Component failure	Political disaster
	Chernobyl	1986	Radiation	31 fatalities
	Tokaimura	1999	Radiation / human error	2 fatalities
Defence	Dhahran	1991	Missile software	28 fatalities
	Chinook ZD576	1994	Human error / Software (?)	29 fatalities
	Osprey Marana	2000	Craft stability / human error	19 fatalities
Construction	Milford Haven	1970	Design flaw	Policy change
	Daegu subway	1995	Gas explosion	101 fatalities
	Toledo Ohio	2004	Anchor procedures	4 fatalities, $280k fine
Aircraft	Kegworth	1989	Component failure / Human error	47 fatalities
	Florida	1996	Oxidiser in hold	110 fatalities
	Concorde	2000	Foreign object	113 fatalities, commercial closure
Space	NASA 51-L	1986	Component failure	7 fatalities
	Arianne 5	1996	Software	Mission loss
	NASA Mars probe	1999	Software	Mission loss
Tourism	Hyatt hotel	1981	Design change flaw	114 fatalities
	Herald of Free Enterprise	1987	Procedural failure	193 fatalities
	Indiana train ride	1996	Component neglect	1 fatality, commercial closure

This compilation of evidence has several names across the many industries and nations of the world, but its focus is always concerned with understanding the safety status of a system with the familiar goal of avoiding future accidents. Some of the titles given (not an exhaustive list) to these processes and documents are as follows:

1. Contemporary Safety Status Report
2. Safety Case & Safety Case Report
3. Annual Safety Report
4. Control of Major Accident Hazards Report
5. Occupational Safety and Health Plan
6. Health and Safety Plan (HASP)
7. Health Hazard Assessment Report
8. System Safety Approach Documentation
9. Safety Assessment Report (SAR).

This book will make reference to many of these, but will inevitably concentrate on just a few as vehicles for discussing the issues relevant to all safety regimes.

The Safety Case

The precise meaning of the term 'safety case' rather depends on your particular relationship with the safety case and the particular purpose the safety case is intended to satisfy. It is likely that each person approaching the phrase 'safety case' will have some preconceived idea about what they are getting involved with. For a safety virgin, this idea is unlikely to be well developed – that is to be expected and is perfectly acceptable. For a seasoned guru or safety 'black-belt' the meaning of 'safety case' will be quite familiar. However, it is of value to review the definitions contemporary with this text so that the readers become familiar with them in general and in the context of the book.

Before approaching the more technical and specialist areas for detailed definitions, it is worth a cursory look through a language dictionary. Mine, published by the Longman Group twenty years ago [Longman 1986] doesn't contain 'safety case' as an entry, I would not expect it to. However, it does contain both 'case' and 'safety'. The combination offers a powerful starting point for a very useful definition.

> Case: n b(1) the evidence supporting a conclusion; b(2) an argument, especially one that is convincing.

> Safety: n 1 the condition of being safe from causing or suffering hurt, injury or loss.

This combination of 'convincing argument and evidence supporting a condition of being safe from hurt, injury or loss' is certainly not trivial. With the addition of a few specific terms for individual areas, this combination from pretty standard dictionary definitions may be seen to be the root of many more complicated and technical descriptions of the subject. Well done Longman.

The most recent available technical definition from a UK military standard [MoD 2004] cites the safety case as being;

> Safety Case: A structured argument, supported by a body of evidence that provides a compelling, comprehensible and valid case that a system is safe for a given application in a given operating environment.

The comparison of the dictionary and military standard statements, with over a twenty year gap, highlights an unexpected (to this author at least) but welcome similarity.

The principle aim of a safety case is to derive and present an argument that the system in question will be acceptably safe in a given context. The concept of a safety case is not industry specific, the system could be from any industry. It just needs to be an entity with boundaries, for example a physical system – an engine, a factory, a weapon or a washing machine; it could be procedural for example an oil production facility, a transport network or an assembly line; or it can even be related to some specific event, for example a sports game, a prototype test flight or the demolition of a building. The safety case should contain all necessary information to enable the safety status of the entity to be determined, and while the structure may remain fairly constant, the status of the particular elements will change over the life of the entity, for example planned analysis will be replaced by the analysis results.

Of course the context is all important – a weapon might be considered completely safe when it is not being fired, but it does have other properties that can cause harm, injury or loss. It may have sharp edges and a pointed front end. It may have a significant mass, so when stationary and on its rack it has significant potential energy and when being transported it will have significant kinetic energy. So a lot more than just the explosive energy needs to be analysed when assessing the safety of a weapon system.

Historical incident

An inert missile system used for trials was being transported around a yard area on its trolley. The trolley was being pushed by two persons between store houses at walking pace. The new housing had a lip at the door to allow secure sealing, so the trolley had to be gently 'bumped' over the lip. The front wheels were bumped by person one at the front, who then walked into the store guiding the missile trolley and keeping it straight. Person two bumped the rear trolley wheels, but had to give a significant shove to get the trolley in. The extra effort pushed the trolley towards the back wall of the store and person one instinctively attempted to stop the trolley with his hand. The hand was crushed between the trolley and the back wall.

This manual handling procedure had been reviewed and designed with safety in mind. Transportation was done at walking pace with two persons for maximum control. The trolley was specifically designed for the weapon system in use so that the missile could not be dropped or worked free. It was considered very safe. However, the interaction with the storage system was not considered – the store was not considered to be part of the weapon system, and was not considered to be part of the transportation process. The boundary for safety analysis was set too small, the context was not wide enough.

The Safety Case Report

The safety case is the whole safety justification – just as is a case for law, it comprises every appropriate piece of evidence to make a convincing argument to support some conclusion about guilt or innocence. In this case the argument concerns the safety performance of some entity or system. As a collection of evidence it needs a guide to describe how the evidence was obtained, why it was obtained and what deductions can be made from it. In a court of law, this is done by the solicitor or attorney, but in a safety case this is done by the safety engineer through the safety case report. This report summarises all the key component parts of the safety case, it makes the safety argument explicit and describes the supporting evidence. All supporting documents, analysis and results should be referenced from the safety case report. This evidence does need to be available for scrutiny, but it does not need to bulk out the safety case report.

The safety case report should cite evidence that indicates that the entity, process or system in question meets all applicable legislation and standards. It should confirm that key staff are in place with defined responsibilities; that any further safety requirements and targets that have been set and met are appropriate; that hazard analysis has been carried out correctly; that the level of residual risk is tolerable; and that the safety performance of the entity, process or system has been independently assessed.

Several UK industries have legal obligations to produce a safety case for their operation, for example, rail, nuclear, petrochemical and some other chemical facilities. Several more industries have made the creation and provision of a safety case a mandatory part of satisfying contract conditions, for example the defence industry. Without the safety case, contracts are breached and legal redress is sought. Still more individual companies have adopted safety cases as a 'good idea' to put rigour and process into their safety programmes. The contents, development process and management of safety cases and safety case reports are obviously fundamental topics and will be the subject of later chapters in this book.

Health and Safety Plan

Again, before getting to the more technical descriptions of what a Health and Safety Plan actually is, there should be the customary review of the standard dictionary definitions [Longman 1986]. Not surprisingly, the phrase is not listed on its own, but the individual items are:

> Health: n 2, condition <*of the body*> esp. sound or flourishing; well-being.

> Safety: n 1, the condition of being safe from causing or suffering hurt, injury or loss.

> Plan: n 2, a method for achieving an end, a detailed formulation of a programme of action.

So combining these together leads to a detailed programme of action to achieve the conditions of being safe from suffering hurt, injury or loss, and of flourishing with well-being. Again not bad, perhaps a bit wordy, but it would appear to be perfectly clear and reasonable.

This plan does have different areas of focus in different countries and industries. In the US, the Health and Safety Plan or HASP specifically addresses hazardous waste. This includes decontamination and clean-up of a hazardous waste site and investigating the potential presence of hazardous substances. The key elements of a HASP [DOE 1994], whilst having the specific objectives described above, would be useful in many other safety related planning programmes. They are as follows:

1. Site characterisation and system description
2. Identifying the safety and health risks
3. Specifying requirements for personal protective equipment
4. Specifying requirements for health surveillance
5. Site control, monitoring and decontamination
6. Production of an emergency response plan
7. Procedures for confined entry and spill containment.

An electronic assist is available from US Government websites to give a lead through the development of each of these written elements, and to allow the incorporation of site specific detailed information.

In the UK a Health and Safety Plan again has a specific job function – however, it is very much different from that in the US. The construction industry is the focus for the UK HASP, it is the subject of The Construction (Design and Management) Regulations [HMSO 1994], which aims to improve the management of health, safety and welfare of construction workers through all stages of a construction project. Adherence to the regulations also ensures that critical safety information about a building is available for construction workers and users throughout and after the construction process.

As part of tendering for a construction contract a Health and Safety Plan must be submitted. The pre-tender plan must be developed for the construction phase to include:

1. A full description of the project
2. Arrangements for managing the project
3. Arrangements for monitoring compliance with health and safety requirements

4. The identified risks to health and safety
5. Arrangements for the welfare of people associated with the project.

Upon inspection there is a good comparison between the international uses of the HASP, even though the plans are used for different industries, the objectives and contents are remarkably similar. As with the safety case, the use of the HASP does not necessarily need to be industry specific. The approaches set down would be equally applicable to any industry, any project, and any system.

System Safety Approach Documentation

The System Safety Approach is approved for use by all departments and agencies within the US Department of Defense (DoD) [DoD 2000]. Its objectives are to protect private and public personnel from accidental death, injury or occupational illness; also to protect public property, equipment, weapon systems, material and facilities from accidental destruction or damage while executing missions of national defence. Within mission requirements, the DoD will also ensure that the quality of the environment is protected to the maximum extent that is practical. The scope of the system safety approach covers the management of environment, safety and health mishap risks during the development, testing, production, use and disposal of DoD systems. The forward to the approach standard also notes that the safety goal is zero mishaps.

In common with the introduction to the other approaches to safety, it is again worth referring to dictionary [Longman 1986] definitions of some of the main terms used here.

> System: n 1c A group of interrelated and interdependent objects or units; 2 An organised set of doctrines or principles usually intended to explain the arrangement or working of a whole body.

> Safety: n 1 the condition of being safe from causing or suffering hurt, injury or loss.

> Approach: n 2 A manner or method of doing something, especially for the first time.

Together, these terms give a good description of the intent of a System Safety Approach, but they don't match up to the DoD definitions [DoD 2000], which are as follows:

> System: An integrated composite of people, products, and processes that provide a capability to satisfy a stated need or objective.

> Safety: Freedom from those conditions that can cause death, injury, occupational illness, damage to or loss of equipment or property, or damage to the environment.

There isn't a cited definition for 'Approach', but there is one for 'System Safety' which goes further and has a different focus than the two separate definitions given above. It is as follows;

> System safety: The application of engineering and management principles, criteria, and techniques to achieve acceptable mishap risk, within the constraints of operational effectiveness and suitability, time, and cost, throughout all phases of the system life cycle.

To enable further understanding I would like to draw out the meaning of the word 'mishap'. This may sound rather a quaint term, as if one had tripped over a shoelace, but it actually has a very much more serious meaning than this when used in the context of safety. From my dictionary;

> Mishap: n An unfortunate accident.

and from the DoD [DoD 2000];

> Mishap. An unplanned event or series of events resulting in death, injury, occupational illness, damage to or loss of equipment or property, or damage to the environment.

Overall the DoD system safety approach is sound, although the definitions have to have more thought applied to them to follow them through. On the whole, the definitions do compare well with the earlier defined terms from the UK and Europe, and from the different industry fields. At this stage there is not a consistent term used for the collection of hazard, risk and safety information, but looking behind the varied terms and phrases used, the intent appears to remain largely consistent.

As with the other areas looked at so far, there is a requisite set of documentation of the system safety approach. The objective of this document suite called by the DoD is to record the developer's and program manager's approved system safety engineering approach. The documentation should:

1. Identify each hazard analysis and mishap risk assessment process used
2. Include information on safety integration into the overall program structure
3. Define how hazards and residual mishap risks are communicated to and accepted by the appropriate risk acceptance authority
4. Define how hazards and residual mishap risks will be tracked through the program life.

There are a series of steps and results recording to go through when implementing the systems safety approach. These are described as follows:

1. Identification of hazards
2. Assessment of mishap risks
3. Identification of mishap risk mitigation measures
4. Reduce mishap risk to an acceptable level
5. Verification of mishap risk reduction
6. Acceptance of residual mishap risks

7. Track mishap risk throughout the system life cycle.

It is worth a comparative look back at some of the steps previously brought out under other safety approaches – there is no single catchy collective phrase for all these processes and evidence, but the consistency is surprising and certainly most welcome. These processes are fundamental topics and will be the subjects of later chapters in this book.

Control of Major Accident Hazards (COMAH)

The main aim of the UK COMAH regulations [HMSO 1999] is to prevent and mitigate the effects of major accidents involving dangerous substances, such as benzene, liquefied petroleum gas, explosives, certain nuclear materials and arsenic pentoxide which can cause serious damage/harm to people and/or the environment. It is worth noting that the COMAH Regulations treat risks to the environment as seriously as those to people. Sites are designated COMAH applicable due to the quantities and type of hazardous materials under their control – there are two tiers of interest, with the top tier being those sites with the highest quantities of dangerous materials. The top tier sites have a significant number of duties to perform, one of which is to summarise their compliance through the preparation and presentation of a COMAH safety report.

A safety report is a document prepared by the operator of the site and its aim is to demonstrate that all measures necessary for the prevention and mitigation of major accidents have been taken. At this point it is certainly worth reviewing the definition of a 'major accident' – what does this actually mean? The particular statute in place under UK law [HMSO 1999] has the following definition;

> Major Accident: An occurrence (including in particular, a major emission, fire or explosion) resulting from uncontrolled developments in the course of the operation of any establishment and leading to serious danger to human health or the environment, immediate or delayed, inside or outside the establishment, and involving one or more dangerous substances.

As good as this is, there are no specifics on how much of an emission of nasty chemicals is major and how much is not. Well, within the regulations, a list of categories and quantities of dangerous substances is given, which the regulations apply to. These are as follows:

1. Very Toxic   5000 Kg
2. Toxic   50000 Kg
3. Oxidising   50000 Kg
4. Explosive   10000 Kg
5. Extremely Flammable   10000 Kg
6. Highly Flammable   20000 Kg
7. Highly Flammable (Liquid)   5000 Tonnes
8. Flammable   5000 Tonnes

9. Dangerous for the Environment   200 Tonnes
10. Material reacts violently with water   50000 Kg

Specific meanings are given in the appendices of the regulations [HMSO 1999] for the differences between the levels of flammability and toxicity. The HSE itself gives an indication of the quantities of these types of materials which have to be involved for a class of major to be called and for official notification to the executive to be mandatory [HSE 1999].

1. Sudden, uncontrolled release in a building of: 100 kg or more of flammable liquid; 10 kg of flammable liquid above its boiling point; 10 kg or more of flammable gas;
2. Sudden, uncontrolled release in a building of 500 kg of these substances if the release is in the open air; and accidental release of any (quantity of any) substance which may damage health.

An another criterion used is the effect on any local population. If the total length of time a population is required to remain indoors or quarantined exceeds 500 person hours (for example 100 people for 5 hours, or 1000 people for half an hour), the incident will still be classed as a major accident even if no-one is actually injured.

Historical Incident

A COMAH top tier establishment produces a range of chemicals including motor fuel additives, chlorine and solvents. It is top tier because of the inventory of lead alkyls, chlorine, liquefied extremely flammable gases and other toxic chemicals. On Sunday 11 July 1999, a road tanker containing 20 tonnes of molten sodium had been returned from a customer and was being heated to melt the sodium prior to unloading. This caused a positive pressure within the vessel. The operators failed to vent the pressure as per standard operating procedures. Sodium had solidified in the outlet valve and a plant operator attempted to clear it using a metal rod. When he did so, 4 tonnes (1,800 lbs.) of molten sodium spilled out and ignited. The on-site and off-site emergency plans were activated. The on-site emergency response team succeeded in putting the fire out after 3 hours, by smothering it with sand. The police instructed local residents to remain indoors and more than 1000 people were confined to their homes for 3 hours. The nearby M53 motorway was closed for 45 minutes and a local charity football match disrupted. This is a major accident because the confinement of people indoors exceeded 500 person hours. There were no injuries or off-site damage. The cause was operator error, in failing to follow the correct operating procedures for clearing a blockage in the road tanker outlet. The company had to demolish the offloading facility and rebuild to modern standards at a cost of £200,000.[HSE 2001]

Summary

There are multiple requirements throughout the world for risk and safety analysis in a wide variety of industries. It is unfortunate that they are all identified by different terms and phrases, and it is not the specific aim of this book to say that any descriptive term is better than any other. The main interest of citing the definitions and objectives used, is to demonstrate that even though different words are used, there is a hugely significant overlap in process and goal.

This will mean that as you go through the rest of this text, even though the content may be discussing some part of a safety case or a system safety approach, you should feel reasonably comfortable that the discussion is not confined or limited to that particular safety documentation.

Notes

DoD 2000: "Standard Practice for System Safety" Military Standard 882D, United States Department of Defence, February 2000.

DOE 1994: "Hazard Baseline Documentation" DOE-EM-STD-5502-94, Section 5.5. United States Department of Energy, August 1994.

HMSO 1994: "The Construction (Design and Management) Regulations" Statutory Instrument 1994 No. 3140. Her Majesty's Stationary Office, London, 1994.

HMSO 1999: "The Control of Major Accident Hazards (COMAH) Regulations" Statutory Instrument 1999 No. 743. Her Majesty's Stationary Office, London, 1999.

HSE 1999: "Explained – Reporting of Injuries, Diseases and Dangerous Occurrences Regulations" Leaflet HSE31(rev1). The Health and Safety Executive, June 2004.

HSE 2001: "Major Accidents Notified to the European Commission for England, Wales and Scotland 1999-2000" The Health and Safety Executive, October 2001.

Longman 1986: "English Dictionary and Roget's Thesaurus" Longman Group UK Ltd, 1986.

MoD 2004: "Safety Management Requirements for Defence Systems Part 1" Interim Defence Standard 00:56, Issue 3. Ministry of Defence, December 2004.

Chapter Two

The Language of Safety

The Concepts of Language

All language has the same goal – to communicate someone's thoughts. When those thoughts are about matters of life and death (as in safety work), it is vital that the writer and reader have the same understanding of the original thoughts. Modern English language has a great ability to express many things in many different ways. It sometimes seems that there is a word for everything we could possibly want to express. Unfortunately this is not the case. We do have words for the graded separations of 'high', 'medium' and 'low', but we only have 'safe' and 'unsafe'. There is no single word for the level of safety that is at an acceptable level between safe and unsafe. Perhaps, there isn't one, that is, the concept of safety is a binary condition, you either are or are not – there is no in between. This debate is on going, and I do not intend to settle it myself. This text will consider that there is a mid-region between safe and unsafe, even if the region is just thought of as a 'line', rather than a continental sized area.

Within English as a language there is a problem. It's a process called *ellipsis* and it allows you to leave out words you think are obvious, and it is perfectly acceptable grammar. For example, if I tell you that I am presenting some equipment safety evidence in a report, and you ask "What has been demonstrated?", you actually mean 'What has been demonstrated *about the equipment by the safety evidence* ?' You just left out the second half of the sentence. This is perfectly fine because you and I are both fully aware of the context of our statements. But I have seen examples in real safety documents where after a while, I'm not sure about what was trying to be said. This is because of unconscious use of the ellipsis process. Here is a paragraph of text from a real safety requirement document introducing the report section on the tracking of software failures;

> **Visible bug tracking**
> Here we provide evidence of bug tracking for the software. 'XXXXX' is the database that is used to track all issues regarding this system. It has full visibility and is extremely detailed.

You might think this is perfectly understandable, but there are a number of areas where extra context has been assumed, and so I am left thinking have I really understood what has been written. What is actually meant by 'all issues'? Should this really be 'all *software* issues', 'all *bug* issues' or 'all *safety* issues'? What does 'full visibility' mean? Full visibility of what? 'Full visibility *of the software'*? 'Full visibility *of bug information'*? Does the 'it' really mean that 'it' *presents* full visibility to the viewer? Or is there something more here, perhaps some extra functionality that we need to know more about? From the text as it is, we just don't know, we have to make assumptions. And as you should already know making assumptions can make an 'ass' out of 'u' and 'me'.

Great caution is urged when writing safety reports, please try to be aware of the use of the ellipsis process. Putting in more words does make the written report more voluminous and therefore potentially less likely to be read extensively by a busy superior, but leaving out words can cause confusion, delay and be very dangerous.

This point has led onto several graphical based tools being used to paint a picture of the safety status, reasoning and strategies employed. Later chapters of this book will consider these in more detail.

The Language of Risk, Chance, Probability and Hazard

These words are very often used interchangeably but actually have very different meanings. Each of the words have certainly been used and swapped around in the safety reports that I have seen. Consider the following three statements:

> There is a risk of an accident.
> There is a chance of an accident.
> There is a probability of an accident.

As the reader of these statements, ask yourself if is there any (real) difference in these three statements, from your current understanding of what risk, chance and probability mean. One difference might be the potential numerical relationship between these non-numerate terms. *Risk* has the idea of something that might really happen, *chance* seems to imply that the something might not happen and *probability* certainly has the message that something will definitely happen. However, this is subjective and certainly not consistent across all safety practitioners. It is always worth a look at how dictionaries [Longman 1986] have defined these terms.

> risk: n 1 possibility of loss, injury or damage. 2 a dangerous element or factor; a hazard.

> risk: vt 1 to expose to hazard or danger.

> chance: n 1a an event without discernible human intention or observable cause ... 5 a risk.

> chance:vt to accept the hazard of; risk.

> hazard: n 1 something that may be dangerous.

hazard: n2 a risk that cannot be avoided.

hazard: v2 to risk losing your money, property etc in an attempt to gain something.

probability: n 3 a measure of the likelihood that a given event will occur, usually expressed as the ratio of the number of times the event occurs in a test series, to the total number of trials in the series.'

There are a number of points worth discussing on these definitions. *Probability* is definitely linked to numbers; *risk* nearly gets there with the use of *possibility*. *Risk* certainly introduces the idea of something negative happening. *Chance,* on the other hand, appears to be either good or bad. The interchangeability between *risk* and *chance* is given acceptability by the transitive verb (vt) definition of *chance* being the acceptance of *risk*.

Most of the time, it is some definition of probability (akin to selecting a particular playing card) that people think they are referring to when they use all the terms. However, they are different words and do have different meanings, even if in our modern language use they are often used interchangeably to mean the same thing. As cited at the start of the chapter, it is essential to have a shared meaning between author and reader, and it the responsibility of the author to do this, because they are her or his thoughts that are being expressed.

The Origins of Chance, Risk and Probability

Chance originally meant 'that which befalls', it has come to modern language use from a Latin root 'cadere' meaning 'fall' and via Old French as the word 'cheoir' whose noun derivative includes 'chaunce' [Ayto 1990]. Chance has a 'God's will' aspect to it, as if the person involved has no influence over the outcome, perhaps as in an earthquake. There seems to be a completely random element to chance – that is perhaps why dice and cards are sometimes referred to as games of chance. Although, having stated that, there is always the option of not taking part in the game at all, one of the classic arguments for safety.

The ultimate origins of 'risk' have not yet been satisfactorily explained. Its recent history is sure, English acquired it via the French word 'risque' and the Italian 'risco', which is a derivative of the word 'riscare' meaning 'to run into danger'. One potential origin might be related to an earlier meaning of 'sail into danger' perhaps by sailing too close to rocks ('rhiza' being Greek for cliff), but this has yet to be proven [Ayto 1990]. One thing is obvious however, and it is the difference between chance and risk – at least from their origins, risk is something the person involved has chosen to do. Running or sailing into danger contains a positive decision to accept the danger involved, to go and run the risk. This would be in return for some benefit – getting to port quicker for example.

The concept of deciding to take a risk for some benefit does carry over into modern use of the word in the safety domain. In modern safety language risk is taken as having two parts to it – the idea of some severity of impact (the level of nastiness of the event consequences), and the likely probability that the impact will occur. These two factors must be treated in combination to have a full understanding of risk. As the risk of fatality is often the most severe impact in the safety domain, the term 'risk of fatality (per unit)' often gets shortened to just 'risk' via use of the ellipsis process. It is always worth posing the questions; risk of what? and how often? It may be seen as annoying by others (the voice of experience here), but it can often lead to a valuable review of just what it is that everyone is discussing.

Probability derives from Latin 'probabilis' and the Middle French word 'probare' meaning to test, approve and prove. The further French word 'probus' means good and honest [Ayto 1990]. On the face of it, it doesn't seem to have any mathematical origins. The mathematical significance for probability became more important when gambling was (even more) popular. A gentleman in France called The Chevalier de Mere invited mathematician Blaise Pascal to carry out testing and proving on a gambling problem he was considering. De Mere played a game in which one would bet on the likelihood of throwing a six during four throws of a die. This progressed onto the likelihood of throwing a double six during 24 throws. Gamblers of the time believed that the two games were equally fair as the ratio of possible outcomes to number of throws was the same (4 to 6 for a single die, and 24 to 36 for two dice), giving a break even ratio of 0.666. It was shown that the single die game gave the probability of throwing a six from four throws as just over a half, and so was balanced in the favour of the thrower. In the two dice game, the probability of throwing a double six within 24 throws was shown to be just under a half, and so was balanced against the thrower. One would need to throw the two dice one more time to get a balance in favour of the thrower. The relationship of the break even ratio was eventually solved by a man called Abraham de Moivre in 1716, the factor he calculated was 0.693 – which interestingly turns out to be the natural logarithm of 2. The actual expression for carrying out the calculation is;

Throws to obtain a specific result = ln2 x Number of possible outcomes

So, the number of throws needed in order to have at least a 0.5 likelihood of throwing at least one double six is 0.693 x 36 = 24.95. Much nearer to 25 than 24 [Webb 1996].

The Origins of Hazard

Back before the Middle Ages, the North Africans played a gambling game using little cubes with numbers, called 'az-zahr' the Arab word for the die itself. The game transferred across the Mediterranean to France, where it was named 'Hasard', it then moved over the English Channel to England some time before 1500 AD where it was given the English spelling of the same word, 'Hazard'. The word came to mean an unlucky throw of the dice, since you would lose your bet if the throw came out incorrectly.

This word now has the meaning in the safety domain of a source of danger, or a situation with potential to cause harm. In recent history hazards are always there, it is the control of them that prevents an accident from occurring. This was not the original meaning of the word, which was more akin to chance i.e. something you could not avoid and had little control over the result of playing the game (of chance). Incidentally, that particular game is still played in modern casinos – it's now called 'Craps'.

The differences in origin and meaning of these four terms, especially 'risk' where you may have a decision to make, and 'chance' and 'hazard' where you may not, are critical to their accurate use in modern safety language and must be mutually understood when discussing situations in this domain. Please take a few moments to recall your last use of one of these words, and ask yourself if you really meant what you said, or said what you really meant.

The Origins of Safety and Safety Case

Safety derived from the Latin 'salvus' meaning 'uninjured'. The same root has also given us 'salvage', 'salvation' and 'solid', all of which are obviously related to being rescued and sound. Again, it reached English from the French via the word 'sauf'. Salvus itself goes back even further to ancient Indo-European languages with 'solwos' meaning 'whole'. Another derivative of salvus has led to the herb name 'sage', which has the etymological definition of 'healing or saving plant', due to its medicinal properties [Ayto 1990].

The concept of safety has been around and understood for thousands of years. The English word 'safety' has probably been around for hundreds of years. The concept of a safety case and safety reporting has probably been around for only a few tens of years. We have already discussed several meanings and definitions of the phrase safety case, but have yet to look at the origins of its use.

The earliest legislative requirement for a safety case in the UK come from The Nuclear Installations Act of 1965 section 14 [HMSO 1965] on safety documentation states;

> Without prejudice to any other requirements of the conditions attached to this licence the licensee shall make and implement adequate arrangements for the production and assessment of safety cases consisting of documentation to justify safety during the design, construction, manufacture, commissioning, operation and decommissioning phases of the installation.

This document also gives the earliest definition and purpose of a safety case – 'to justify safety', and some direction for the focus of the justification, '… during design, construction, manufacture, commissioning and decommissioning phases of the installation.' Wow!! This is pretty comprehensive and has been the forerunner of many of the UK standards and requirements for safety cases.

Modern use of Safety Language

The UK Engineering Council has published a set of guidelines on risk issues [Engineering Council 1993], this is not a particularly new document, but it does contain significant information on public awareness of risk. It cites that '... engineers should learn about how the public perceives risk and makes risk decisions. Conversely, engineers should inform the public about how the profession perceives risks and makes risk decisions'. Modern use of safety language is about communicating about risk and safety. This is most important for governments and policy developers. Notably, the UK and Australian governments have supported research into how the public perceives risk and safety issues [Cabinet Office 2002], [Botterill & Mazur 2004]. These have suggested that the language of risk is used to cover a wide range of types of issues:

1. Direct threats from terrorism
2. Safety issues (BSE, MMR, flooding)
3. Risks to the environment
4. Transfer of risk to and from the private sector
5. Risk of damage to a government's reputation.

The language itself can also be confusing. People often give different meanings to key terms so it is important to develop a commonly understood safety language, which should be capable of being understood by those outside as well as inside government [Cabinet Office 2002]. These comments are from the UK government, and the Australian government is in similar agreement. Risk is central to policy response to drought and quarantine restrictions; terms like 'risk management' and 'acceptable levels of protection' assume a degree of understanding of the concept of risk, risk acceptance and how risk is measured. These are bold assumptions. Understanding how stakeholders and the broader community perceive risk will assist policy makers in developing better policy and more effective means for communicating in areas involving risk and safety management [Botterill & Mazur 2004].

Development of the Safety Case in the UK

In the UK the legislative requirement for a safety case has moved through various hazardous employment fields. Usually this has come about after a public inquiry into some dreadful accident in each field. The timetable of the safety case's progress through UK industry is given in table 2.1 – details of the incident generally recognised as the justification are also given.

The Aircraft and Armament Evaluation Establishment (A&AEE) document cited in table 2.1 [A&AEE 1992] gives a further brief overview of a definition of the purpose of the safety case:

a. Identify the potential hazards that could arise
b. Categorise the effects of those hazards
c. Quantify the probability of those hazards
d. Justify the acceptance of those hazards or identify design changes needed
e. Provide a permanent record of all the above to be updated through life.

It should be noted that the emphasis in the safety case descriptions so far is based on prediction of future behaviour and identifying the potential hazards. This is generally consistent with the on-going use of safety cases in the UK, although there is sometimes the need for the production of a retrospective safety case in some situations.

Table 2.1 The Progress of the Safety Case in the UK

Industry	*Incident*	*Legislation / Requirements*
Nuclear	Windscale 1957. Fuel fire as a result of errors during Wigner energy release. No direct fatalities, but substantial release of radioactive material.	Nuclear Installations Act 1965
Chemical	Flixborough 1974. Uncontrolled modifications to a chemical process line led to the pipeline rupturing and a huge explosion killing 27 people.	Control of Industrial Major Hazards (CIMAH) 1984; and Control of Major Accident Hazards (COMAH) 1999.
Rail Transport	Kings Cross 1987. Fire started underneath an escalator, which rapidly engulfed the main exit routes. 31 people killed.	Railways (Safety Case) Regulations 2000.
Petrochemical	Piper Alpha 1988. Small initial explosion of condensate pump led to catastrophic fire killing 167 people and costing £2000 million.	Offshore Installations (Safety Case) Regulations 1992.
Defence	Various, probably including; 27 Hawk aircraft lost 1980 to 1989; 23 Tornado aircraft lost 1980 to 1989; 24 Sea King helicopters lost 1980 to 1989.	Outline of Requirements for the Provision of a Safety Case by the Design Authority, A&AEE, 1992; and Safety Management Requirements for Defence Systems 00:56 1991

Development of Safety Reports in the US

In the USA the driving forces for safety have been the nuclear and space programmes. The phrase 'safety case' is not widely used, which may come as a surprise when noting the importance given to litigation in the USA. A variety of phrases and terms are employed, and taking NASA as the example, the NASA Safety Manual Procedural requirements [NASA 2000] cite:

> Safety Analysis Report (SAR). A safety report of considerable detail prepared by or for the program detailing the safety features of a particular [nuclear] system or source.

and,

> Mission Safety Evaluation (MSE) Report. A formal report for a specified mission to document the independent safety evaluation of safety risk factors that represent a change, or potential change, to the risk baseline of the program.

NASA also has involvement in the US Aviation Safety Reporting System (ASRS), which was set up in 1975 under a memorandum of agreement between NASA and the FAA. Its purpose is to collect, analyse and respond to voluntarily submitted historical reports of safety related incidents, with the goal of reducing the likelihood of future aviation accidents. The ASRS specifies an Aviation Safety Report Form and there are different versions depending on your particular role in aviation. The version for pilots is shown on the following pages.

Whilst this is a reporting system based on incidents that have happened, that is, it does not deal with the prediction of incidents, the document layout is a useful example of the sorts of details required in a simple safety report. However, there are some extremely important concepts within it, which will be discussed, in later chapters.

The Occupational Health and Safety Agency (OSHA) is the primary force behind employee safety in the USA. Its main Act of 1970 [Note to UK readers:- This was obviously in advance of the UK Health and Safety at Work Act of 1974], cites several documents for recording safety matters. In the Act, the US Congress declared it to be the Acts' purpose and policy ' ... to assure so far as possible every working man and woman in the Nation safe and healthful working conditions and to preserve our human resources'. It encourages each State to assume the fullest responsibility for the administration and enforcement of their occupational and health laws. In order to accomplish this, each State, if they desire to, may submit a State Plan for the development and enforcement of their occupational health and safety standards.

Safety management has developed over time in the occupational arena to 1989 when OSHA issued recommended guidelines for the effective management and protection of worker safety and health [OSHA 1989]. These are still applied today. In summary the general points are;

Employers are advised and encouraged to institute and maintain in their establishments a (safety) program that provides adequate systematic policies, procedures, and practices to protect their employees from, and allow them to recognize, job-related safety and health hazards.

An effective program includes provisions for the systematic identification, evaluation, and prevention or control of general workplace hazards, specific job hazards, and potential hazards that may arise from foreseeable conditions.

Although compliance with the law, including specific OSHA standards, is an important objective, an effective program looks beyond specific requirements of law to address all hazards. It will seek to prevent injuries and illnesses, whether or not compliance is at issue.

The extent to which the program is described in writing is less important than how effective it is in practice. As the size of a worksite or the complexity of a hazardous operation increases, however, the need for written guidance increases to ensure clear communication of policies and priorities as well as a consistent and fair application of rules.

Summary

It is interesting to note that many of the US OSHA points could easily be referring to a 'safety case' from the UK. Once again, as in the summary to Chapter 1, the underlying concepts are strongly comparable across the nations and do not appear to contrast significantly at all. The predictive aspects are present and the idea of forming a written record is also noted. The only difference it seems, is the label that is attached to the documents.

Notes

A&AEE 1992: "Outline of Requirements for the Provision of a Safety Case by the Design Authority – Reference AEN/18/103", Aircraft & Armament Evaluation Establishment, MoD, February 1992.

Ayto 1993: "Dictionary of Word Origins", John Ayto, Bloomsbury, 1993.

Cotterill & Mazur 2004 "Risk and Risk Perception – A Literature Review", Australian Rural Industries Research and Development Corporation, 2004.

HMSO 1965: "Nuclear Installations Act 1965 – Section 14", Her Majesty's Stationary Office, London, 1965 (reprinted 1993).

Longman 1986: "English Dictionary and Roget's Thesaurus" Longman Group UK Ltd, 1986.

NASA 2000: "NASA Safety Manual w/Change 2, 03/31/04 (NPR) 8715.3", NASA 2000 (re-validated 2004).

OSHA 1989: "Safety and Health Program Management Guidelines 1926

Subpart C", Occupational Safety and Health Administration, US Department of Labor, 1989.

Strategy Unit 2002 "Risk: Improving Government's Capability to Handle Risk and Uncertainty", The Cabinet Office, London, 2002. Crown copyright.

The Engineering Council 1993. "Guidelines on Risk Issues", The Engineering Council, London. pp. 25, 1993.

Webb 1996: "The Layman's Guide to Probability Theory" Peter Webb 1996-2005, at http://www.probabilitytheory.info/

Chapter Three

The Safety Management System

The Components of a Safety Management System

A Safety Management System contains all the items used in managing safety. This must be understood and recorded if an understanding of the safety situation relating to something is to be obtained. This includes all the people, all the procedures, all the hardware and all the computers and software that is employed within the system that has an effect on the level of safety of the system. The safety management process is actually going to deliver the safe functioning of the system. Many of the component parts will be fairly obvious – a simple fire protection system for example:

1. There will be hardware – the extinguishers and sprinkler systems (from water store to sprinkler head)
2. The training and operating procedures for using the extinguishers, raising the alarm and undertaking an evacuation
3. There may well be fire control officers and certainly fire-fighters even if they are the external emergency services
4. There may also be smoke, heat and infra-red detectors relying on software and computers.

This is just for a simple fire protection system in an office building, imagine what the fire safety management system would be like for an entire offshore oil installation. This is just one component of the overall Safety Management System – there will be similar system components for all the safety risks present on a particular site or within a particular operation.

One of the contributory causes to many of the publicly known accidents (as cited in Chapter 1) has been shown to be management failure. This is not always the 'Management' as a group of people, this is 'management' as a corporate function – for which all employees have some responsibility. It is absolutely true that those in senior management positions should take a lead in safety – otherwise how can the employees lower down the pay scales be expected to understand their own responsibilities.

As an excellent example of this NASA's Safety Manual [NASA 2004] states in Chapter 1, Part 1, Section 1.1.1 (i.e. right at the start!) that:

> Safety program responsibility starts at the top with senior management's role of developing policies, providing strategies and resources, and is executed by the immediate task supervisor and line organization. All employees are responsible

for their own safety, as well as that of others whom their actions may affect (Requirement 25001).

In many industries, the lack of this type of management lead within the safety system has, and no doubt will still, lead to tragic accidents.

Historical Incident

The 1987 King's Cross underground fire, probably started by a discarded cigarette even though there was a smoking ban in place, quickly propagated to engulf an escalator filled with members of the public. At the public enquiry, the contributory causes were highlighted as maintenance of equipment was not performed properly; the lack of formal risk assessment; no evacuation procedures; no rehearsals had taken place (that would have highlighted that emergency radios would not work underground). The management team was lulled into a false sense of security by the fact that no previous escalator fire had caused a fatality. No one person was charged with the overall responsibility and management of safety.

Engineers and scientists in managerial positions should recognise that they are likely to have enhanced responsibilities in several safety related areas [Engineering Council 1993], for example:

1. The introduction and operation of a working safety management system
2. The discharge of their employee duties, so that they do not become a source of risks to safety
3. The responsibility of making judgements relating to the tolerability of risk.

The whole 'Corporate' safety management scope should also reflect this approach. The management and approach to safety should be as systematic, planned and focussed as the effort applied to any other critical business process. A life simply cannot be recovered in the next fiscal quarter.

Designing a Safety Management System

It will probably be likely that you or your company already have some concept of managing safety. You may not explicitly recognise it as a Safety Management System as such, but there will be some effort made towards keeping people safe. There are a number of key concepts to designing and implementing a satisfactory Safety Management System, each element is structured on the following stages of management [MoD 1996];

- Policy: What are the requirements and objectives for the Safety Management System?

- Organisation: Who is responsible for delivering the policy?
- Implementation: How is the policy is to be delivered?
- Measuring Performance: What are the arrangements for monitoring the system behaviour?
- Review and Development: How will past performance be incorporated into future improvements?

Understanding the components of a system that have an influence on safety is critical to having a low risk operation. Thinking about the people, procedures, hardware and software as inter-operating objects is a valuable process to understand and organise the way a safety report is presented. The object descriptions and relevant stages of management should be recorded in a written form – the precise name of the document is not so important, it may simply be called the Safety Management System Document! So for our example of the fire control system, this document may look something like this.

Policy: The fire control system is designed to reduce the risk of fire propagation and to allow evacuation of personnel to safe areas.

Organisation: The Managing Director has overall responsibility for safe operation of the organisational structure to deliver the fire control system policy. The fire safety manager is responsible for implementing a fire control system.

Implementation: The fire control system will comprise a combination of extinguishers and fixed sprinklers; smoke and heat detectors with audible fire alarms; procedures for testing all the appliances; training in the use procedures of the extinguishers; evacuation planning and exercises; and a reporting procedure for capturing the records of the implementation.

Measuring: Annual testing of the sprinkler system; monthly testing of the smoke and heat detectors, weekly testing of the audible fire alarms; annual extinguisher-use refresher courses; quarterly fire evacuation exercises. Measurements are pass/fail criteria, completion of training courses and timing of evacuation.

Review: The fire safety manager will report to the board of directors on a quarterly basis, highlighting the performance measurements. The safety report should make recommendations for future development.

This type of Safety Management System document should be produced for every part of the safety features of the system. This should not be viewed as a trivial task as there may be something approaching 100 different entities in a complex system, many of which will inter-relate with other parts of the system. In the imaginary office, there may only be a few, but there will also be a need to review the ergonomics of working position design, use of office equipment (photo-copiers, guillotines and printers), and even cleaning arrangements for windows, desks and floors (what chemicals are to be used and how are they stored?).

For the more complex operation of a petrochemical plant, the safety management document set is likely to be considerable, the effort will be likewise, but then so will be the value of the information.

Safety Management Planning

At any stage of a project or operation there is a requisite set of plans to be produced, reviewed and updated – resource plans, cashflow plans, marketing plans, delivery plans. It is also essential to consider a safety plan detailing how safety is going to be managed through the project. The typical UK safety management plan also has to consider the safety of the natural world, and so is sometimes called 'The safety and environmental management plan'.

There are a number of useful descriptions of what a safety plan should contain and the format it should take. A few will be presented here to demonstrate the principle and it will be seen that although these have been developed by separate bodies for different purposes, even in separate countries, the construct and intent are largely consistent. This is reassuring to note, because it gives encouragement that as a safety community, these different sources have focussed in and recorded the main significant areas of concern in the field of safety planning.

Example of a UK Safety Plan

From the UK Ministry of Defence [MoD 1996] the following advice is given for a project safety management plan– where the text is perhaps topic focussed, I have provided a more generic interpretation.

> Structure. The following structure [and content description] may be adopted as a basis for the safety programme plan:
>
> Part 1 Introduction. This part should describe the system of interest, the project scope and objectives and a brief overview of the way safety will have an effect on the project.
>
> Part 2 Safety requirements. This part should list out all the safety requirements for the system of interest. These requirements will come from legislation, standards and codes of practice. The main purpose of this section is to provide a reference for all the project staff and to act as a record of all the requirements that are intended to be satisfied. This section should also record any interpretation of the requirements or any tailoring [selective adoption or rejection of specific requirements] that has occurred.
>
> Part 3 Management and Control. This part should contain a description of the 'who', 'when' and 'how' parts of implementing the safety plan. It should specifically include the timing of various assessments and reviews; the human

resource structure for the safety programme – including the identification of the main safety personnel and their training requirements; how any sub-contractors are going to be managed; and how records are going to be kept specifically including the records of hazards and safety decisions taken.

Part 4 Analysis and Assessment. This part should describe the safety analyses and assessments that are going to be conducted on the system. It needs to explain what safety information is going to be obtained from each particular analysis, and how that will fit into the whole system safety programme.

Example of a US Safety Plan

From the US Occupational Health and Safety Administration [OSHA 2006], the following advice is given in the Hazardous Waste Operations and Emergency Response Standard (HAZWOPER). Again, where the text has become specific, a more generic interpretation is given.

Employers shall develop and implement a written safety and health plan for their employees involved in hazardous waste operations. The program shall be designed to identify, evaluate and control safety and health hazards, and provide for emergency response during hazardous waste operations. The written program shall incorporate the following.

(A) An organizational structure. This part of the program shall establish the specific chain of command and identify the responsibilities of supervisors and employees.

(B) A comprehensive workplan. This part of the program shall address the tasks and objectives of the system's operations including the processes, logistics and resources required. This shall also include the anticipated clean-up activities as well as normal operations.

(C) A site specific safety and health plan. This part of the program shall address the safety and health hazards specific to each phase of system operation. It shall consist of a hazard and risk analysis for each task; details of personal protective equipment to be used; frequency of contamination monitoring; any site control measures; details of the emergency response plan; and accident control measures.

(D) The safety and health training program. This part of the program shall record the required training, which should thoroughly cover the following areas: Identification of the personnel responsible for safety and health; educating the employees about all hazards within the site or system; training in the use of the protective equipment; specifying the work practices by which an employee can minimize the risks from hazards; training in the safe use of equipment and controls; and having a medical surveillance program in place.

[NOTE: This particular standard specifies that general site workers should receive a minimum of 40 hours of off-site instruction and three days field experience on-site, under the direct supervision of a trained and experienced supervisor. In the author's mind, it remains debatable that training time alone is a reasonable factor in deciding on personnel's suitability for safety work.]

(E) The medical surveillance program. This part of the program shall specify the coverage [personnel to be included], nature and frequency of any medical surveillance program. This should relate to the nature and type of hazard present on the site or within the system e.g. chemical poisons, high noise levels and rotating equipment. The scope and role of the medical assessor should also be confirmed.

(F) The employer's standard operating procedures for safety and health. This part of the program shall address the engineering controls, work practices and protective equipment procedures for the hazards on the site or within the system. Specific attention should be given to selection of protective equipment; manual handling (especially drums and containers); transportation of hazardous materials; confined space entry; and decontamination procedures.

In reviewing these two approaches, a number of key aspects are notably common. The specification of a chain of responsibility is very welcome, and although everyone has a duty towards safety, it really does focus the mind when YOU are specifically delegated some responsibility for safety. The safety, risk and hazard analysis sections are equally comparable, as is the description of the process, site or system of interest. The acknowledgement of safety training is excellent, as is the specification of a medical surveillance program in the US plan. A more generic step noted in both approaches is the health monitoring of the system of interest – the safety review process and how this is to be conducted. Usually this is through that good old business tool – meetings.

Safety Planning Meetings

These will inevitably take place, they are bound to and at some point you will probably have to go to one. You may even be 'fortunate' enough to have to organise or run one. However, the point that they will *certainly* take place, is the very reason why they should be well organised and controlled right from the start – right from the project planning stage.

There are many descriptions of these meetings – most of which are not suitable for this type of book, but it is useful to have a label by which to recognise the forum for reporting, discussing and deciding about safety issues. The meeting may be called 'The Safety Panel', 'The Project Safety Committee', or perhaps 'The Project Safety Working Group'. There is no consistent, mandated term for the meeting or the delegates, so I do not intend to force one on you. However, some aspects that are usually mandated are: that there is a group or team that has responsibility for safety, that they meet on a regular basis and that they actually review, monitor and control the safety performance of the project, operation or working practice. For a significantly sized project it may be worth having a series of meetings at different managerial levels, involving specific domains or managerial levels. For example a Safety Panel for high level policy setting and decision making coupled with a Safety Working Group who carry out and monitor the risk and hazard analysis tasks.

At the planning stage of a project, it is worth the effort to get these meetings organised well and to confirm the delegates of each group. The delegates will vary over the project life through natural staff turnover, and also by function as different aspect of safety come to the fore during a project. A series of terms of reference should be developed to give structure to the scope and function of the meetings. The terms of reference for any meeting should address several key areas: purpose, tasks, responsibilities, membership and how the group will communicate with other stakeholders.

Governmental departments in many countries follow this course of progression and have produced guideline terms of reference for safety meetings. An example is given below for a typical Safety Management Committee [MoD 1996].

Purpose:
> To co-ordinate the efforts of the team and to assist the team leader on safety issues relating to systems and equipment managed by them.

Tasks:
> Prepare and monitor a programme for the preparation and review of safety documentation.
> Advise and assist other teams who have interfacing equipment or systems.
> Set up and keep under review the safety policy and strategy
> Set the thresholds for tolerable and intolerable safety performance, and to provide definitions and limitations within safety planning documents.
> Ensure that other agencies are aware of their safety responsibilities as a consequence of safety boundary definitions.
> Review hazard assessments, risk classifications and recommendations to users.
> Set up and manage an audit programme.
> Plan and direct the training of team personnel in safety management issues.
> Review safety progress and the actions taken to resolve new hazards.
> Provide operator advice on the implementation of control measures.
> Ensure the publication of safety documents is under configuration control
> Monitor the safety performance, which is being achieved in the field.

Responsibility:
> Compile meeting minutes and safety report following each meeting.
> Report to Directors (higher managers) quarterly, or as appropriate.

Membership:
> Team leader
> Safety Manager (if separate)
> Designer
> Customer
> Training authority
> Maintenance authority
> Safety auditor
> Specialist advisors (if required)

Having some terms of reference helps all those concerned to have a more calm approach to safety management and to safety issues in general. The terms provide guidance to the meeting itself and help everyone understand the scope and jurisdiction of the safety management team and the safety management system as a whole.

Notes

NASA 2004, "NASA safety manual w/change 2" Procedural Requirements NPR 8715.3 03/31/04.

The Engineering Council 1993, "Guidelines on Risk Issues", The Engineering Council, London, 1993

MoD 1996, "Safety Management Requirements for Defence Systems", Defence Standard 00:56 Issue 2, December 1996.

OSHA 2006, "Hazardous Waste Operation and Emergency Response (HAZWOPER)" Standard, 29 CFR 1910.120, OSHA April 2006

Chapter Four

The Purpose of a Safety Case

Why Are You Constructing a Safety Case?

Before any safety case is attempted, the rationale and purpose of it must be clearly understood. This is vitally important, because if the specific requirements for compiling a safety case and writing the safety case report are not clear, then the following safety case will also be not clear. At best, a poor safety case will be of little use to whatever it was intended to apply to. It will remain a permanent record of your engineering style and capability – and if the safety work is poorly defined and poorly implemented, few customers are likely to remain in contact with you. At worst, you (personally as well as corporately) may have accidents that could actually have been forecast and prevented. Depending on your industry, you may therefore come under scrutiny from your regulators and other legislative bodies. Criminal charges could follow a stressful investigation. In some cases around the world, businesses have been forced to close.

Many of the international standards and legislation relating to safety cases have an implied purpose attached to them. For example, in order to obtain permission to operate a nuclear power plant in the UK, you must develop a safety case and produce a safety case report describing it [HMSO 1965]. If you want the Australian Ministry of Defence to purchase your latest whiz-bang weapon, you will need to provide a safety case [AGDoD 1998]. These are legislative and procedural specifications that state that you shall provide a safety case. If you don't have one, then you will not be able to operate in that market sector.

Of course most of us don't operate nuclear power plants, or sell to the Australian Government. But there are still reasons for developing safety cases.

The Safety Case as a Record of Residual Risk

For the consequences of some event, action or process to be defined as unsafe, there is some residual risk. If there were no risks at all, then there would be no reason for being concerned about safety – and you could close this book now and put it back on the shelf.

The more likely case is that there is some level of risk in the course of action that is being considered. Developing a safety case through analysis of the risk is one way of finding out details about the level of risk that people or things are being exposed to. It can be very important to understand how much risk there is in carrying out a given process or some task or action. In various countries there are legal limits of risk exposure, above which, there can be prosecutions, even if there hasn't been an accident!

The following purpose for having a safety case comes from the public inquiry into the Piper Alpha Oil Platform disaster [Cullen 1990]. Here it was stated that:

> Primarily the safety case is a matter of ensuring that every company produces a formal safety assessment to assure itself that its operations are safe.

For many of the safety cases I have been involved with, these purposes were the primary reason for the safety project; to understand how much risk there was, and to check to see if the risk level was legally acceptable. This is a noble and ethical purpose in its own right, and will often be accomplished anyway, even if there are other purposes, as noted below, that actually spark the requirement for a safety case.

Safety Cases as a Management Tool During Change

At some point during many projects things will change. New things become old, old things break down, personnel with experience retire (no direct similarity intended), and customers and markets are lost or gained. All these initiate change, and when projects or systems need to change, it is essential that the new one does not result in a net increase in risk.

If some piece of equipment in a system is being replaced, it is perfectly natural to want to know if the new system is going to be as safe as the old one. However, you may find yourself in a position where you don't know the safety performance of the old piece of equipment. This may be because it was never established at the time of its purchase, and/or the purpose of any safety case developed did not include analysis at a level that identified individual equipment types or human tasks.

This is one of the main reasons for thinking about the purpose of the safety case, in relation to change. The fidelity of the safety analysis needs to be comparable to the fidelity of the items, people or system components that are likely to change.

Safety Cases as a Record of Engineering Practice

All the safety decisions, even at an early time in the project, must be traceable - if anything goes wrong in the future, the reasons why something was done in a particular way will be asked for. These decisions should be documented and accepted by the appropriate project authority, perhaps through the workings of a dedicated project safety group. The safety case can act as an audit trail through the safety work of a project. If considered in advance, parts of the evidence set can be deliberately utilised as a record or log of how risks were judged, and what was done about them if they were deemed too high. In some safety related domains this may be termed 'The Hazard Log', and this will be discussed later in the book.

Safety Cases as a Tool in a Court of Law

Because the safety case can act as a record of engineering practice in the safety domain, it can be used to justify the reasons why certain things were and were not done. The safety case can show how the safety management system worked and the authority it had. It can be used as a vehicle for demonstrating the residual risks and hazards that are believed to be present in a system or process. It can show what changes have been made to the system or process to make it safer.

In certain fields of engineering in many countries, there is legislation that mandates that you need to have a safety plan, safety case or safety and health report. In many other fields of business this is not a legal requirement. However, most countries do have some requirement on every employer or business to maintain a duty-of-care over the people and products that they come into contact with. The safety case may be seen and specifically designed as a way of demonstrating the satisfaction of the duty-of-care, not only to employees but also to regulators and the courts.

One way of considering safety cases is as a case for the defence. In a court of law, the prosecutor and the defence present cases. These cases are built from a claim of innocence or guilt and are argued based on evidence. The parallels to the safety case are useful – they equally scare corporate managers. The main difference with the safety case is that you have the opportunity to construct the case in your own time and whilst you are developing the system. You do not have to be called to court to have to start to prepare your defence case, you can do it now while you have most control.

Safety Cases as a Marketing Tool

Being seen as safe is good – nobody likes a risk, especially one which they know is there, but don't know very much else about it. Customers like to feel secure in their business transactions and in the products or services they procure. So if you, as a commercial enterprise, can enhance the safety of your product or service through providing evidence of good safety performance, then your commercial position will be improved.

Safety as a marketing tool can be powerful – consider your opinion on the performance of different car manufacturers. Some brands deliberately advertise their safety features, and so use safety as an important selling point. However, safety is not sufficient on its own to have a positive effect on the market. It would be no good for a car to have a top speed of 25 mph (40 km/h). OK, that would be 'safer' around town, but not suitable for many of the other purposes for which people purchase cars.

The implicit effect of using safety documents as marketing tools, is that they must be declared to the people and stakeholders you are trying to influence. Within a private contract, this is likely to be a condition of the contract anyway. But where there is no contract, or where the relationship is between a company and the general public, a complete disclosure of safety performance information may not be necessary.

Safety Cases as a Route to Fewer Accidents

As noted above, a safety case and the analysis that takes part in producing one will usually reveal a level of residual risk in separate components of the operation or process. This information is valuable. One of its primary uses is to highlight the more dangerous areas of the operation and process. A risk priority list can be produced showing the areas where risk most needs to be reduced. The high priority items should indicate where things might go wrong in a very bad way, either in magnitude or frequency of occurrence. This can then be used to improve the performance of the more critical areas, which should ultimately lead to a reduction in the frequency and severity of future accidents.

Understand your Particular Purpose(s)

The safety case or safety report *can* show a great deal of things, but only if the scope is designed to do so. This needs to be thought about at the start of the process for developing a safety case. If a particular type of detail is missing from the start, then of course it will not be there when you might need to show that you are a responsible organisation, with a dedication to safety and health of the public and employees etc.

It is very likely that there will be several purposes that the safety work needs to fulfil – compliance with legislation, as a record of (good) engineering practice, identify residual risk and reduce potential accidents. Provided these are known early on in the safety work, there is no reason why they should not all be able to be accomplished. Of course the more purposes that are required to be satisfied, the greater the effort and resource that is likely to be needed, and this does need to be kept in mind. It is highly likely that the future operational scope will be quite different from the original design purpose, so some anticipation is required, even if the most pressing requirement is more limited.

The purpose(s) of the safety case must be clearly defined at the start of the process so that the rest of the safety work has a foundation to grow from. The purposes will also act as part of the defining scope and boundary for the safety case, safety plan and/or the safety report.

Notes

Cullen 1990, "The Public Inquiry into the Piper Alpha Disaster." Lord Cullen HMSO Cm 1310, 1990.

AGDoD 1998, "The Procurement of Computer-based Safety Critical Systems" Def (Aust) 5679, Australian Government Department of Defence, August 1998.

HMSO 1965: "Nuclear Installations Act 1965", Section 14, Her Majesty's Stationary Office, London, 1965 (reprinted 1993).

Chapter Five

The Requirement for a Safety Case

Why Do You Need a Safety Case Anyway?

Where some equipment, activity or task is recognised as being dangerous, it is usual for there to be some rules or advice about its use. This may just be as minor as some operating instructions, but perhaps there are inspectors or even a regulatory body that oversees those particular items or that particular industry. Ultimately this regulator may be the local, state or national government, which passes bills or laws prohibiting or mandating certain actions.

In the UK, the government has mandated the provision of a safety case as a requirement for obtaining a licence to operate in certain high-risk industries. Several other industries have voluntarily followed suit by publishing standards explicitly requiring safety cases to be produced.

In the US, OSHA's plans for reducing workplace fatalities does not mandate the use of safety cases, but it is acknowledged that companies with the best safety records have some sort of formal safety and health programme, and a key to these programmes is the annual safety plan. These plans typically have the reduction of work-related accidents as their goals, and they provide a framework for continuous improvement in safety.

Within Europe as a whole, there is no explicit call for safety cases. However, in certain high risk industries there is Europe-wide legislation explicitly requiring operators of establishments coming under the scope of the relevant directives to establish a major accident prevention policy, and under certain conditions, to establish a safety report, an emergency plan and a safety management system. The safety report component of these requirements may be considered equivalent to a safety case.

Legislation for Safety Cases

Historical incident

In 1976, the Seveso accident happened at a chemical plant in Italy, manufacturing pesticides and herbicides. A dense vapour cloud containing tetrachlorodibenzoparadioxin (TCDD) was released from a reactor, used for the production of trichlorofenol. Commonly known as dioxin, this was a poisonous and carcinogenic by-product of an uncontrolled exothermic reaction. Although no immediate fatalities were reported, kilogram quantities of the substance lethal to man even in microgram doses were widely dispersed which resulted in an immediate contamination of some ten square miles of land and vegetation. More than 600 people had to be evacuated from their homes and as many as 2,000 were treated for dioxin poisoning.

Major accidents in the chemical industry have occurred world-wide. In Europe, the Seveso accident in 1976 in particular prompted the adoption of legislation aimed at the prevention and control of such accidents. In 1982, the first EU Directive 82/501/EEC (so-called Seveso Directive) was adopted. On 9 December 1996, the Seveso Directive was replaced by Council Directive 96/82/EC, so-called Seveso II Directive. This directive was extended by the Directive 2003/105/EC. The Seveso II Directive applies to thousands of industrial establishments across Europe where dangerous substances are present in quantities exceeding the thresholds in the directive [Europa 2005].

In the UK, there are specific requirements laid down in regulations and parliamentary Acts for many industries. There really are hundreds, so only a few are presented here, if you think you are covered by any of these, you should look up the full reference for all the specific information. The Management of Health and Safety at Work Regulation 1999 specifically call, for the employer to carry out and record a risk assessment – an integral part of a safety case, although it stops short of explicitly calling for one. The Construction (Design and Management) Regulations 1994 place duties on the planning supervisor to produce a safety plan and maintain a safety file. The licences for operating a nuclear establishment, a petrochemical establishment or operating on the railways explicitly call for the provision of a safety case. For example; the Railways (Safety Case) Regulations were introduced in 1994 as a means of ensuring that health and safety standards on the railways were maintained post-privatisation. They introduced a 'permissioning' regime requiring all railway operators, as a condition of operation, to prepare a safety case setting out their health and safety arrangements. The defence industry in the UK has followed suit in the requirement for a safety case, and although it is not in legislation, it is in UK defence standards, which can be cited as legal requirements in contracts for the provision of defence equipment and services. The main standard is known as "00:56", and it says [MoD 2006]:

> 9.2 The contractor shall produce a safety case for the system on behalf of the duty holder (person responsible at law for safety) …

> 9.5 The safety case shall contain a structured argument demonstrating that the evidence contained therein is sufficient to show that the system is safe. The (safety) argument shall be commensurate with the potential risk posed by the system and the complexity of the system.

(NB, the use of the term 'shall' has a specific legal meaning in many countries – it is an absolute requirement. Other phrases such as 'will' or 'should' are unlikely to carry sufficient weight in law to force the contractual parties to actually carry out the action.)

The UK interpretation of the EU Seveso directives has taken the form of the

Control of Major Accident Hazards (COMAH) Regulations 1999 (updated 2005). These apply mainly to the chemical industry, but are also applicable to some storage activities, explosives, nuclear sites and any other industry where threshold quantities of dangerous substances identified in the regulations are kept or used. All sites that come under the regulations have to produce a major accident prevention policy (MAPP). This should be a short and simple document setting down what is to be achieved, but it should also include a summary of, and further references to, the safety management system that will be used to put the policy into action. Upper tier COMAH sites (those with the highest quantities of dangerous substances) are further required to produce a safety report. The safety report must include [HSE 2005]:

- A policy on how to prevent and mitigate major accidents
- A management system for implementing that policy
- An effective method for identifying and major accidents that might occur
- Measures to prevent and mitigate major accidents
- Information on the safety precautions built into the plant and equipment when it was designed and constructed
- Details of measures (such as fire-fighting, relief system and filters) to limit the consequences of any major accident that might occur
- Information about the emergency plan for the site, which is also used by the local authority in drawing up an off-site emergency plan.

It is also noted in the regulations that, subject to national security and commercial confidentiality, the safety report will be made available to the general public.

The US Occupational Safety and Health Administration has published a huge quantity of occupational safety and health standards – in particular standard 1910, which currently runs to 1450 individual parts (although a minority are reserved and hence empty). Part 119 of this standard [OSHA 1996] concerns process safety management of highly hazardous chemicals. Included in this standard are specific citations that call for:

> 1910.119(d) ... the employer shall complete a compilation of *written process safety information* before conducting any process hazard analysis required by the standard. ... This process safety information shall include information pertaining to the hazards of the highly hazardous chemicals used or produced by the process, information pertaining to the technology of the process, and information pertaining to the equipment in the process.

> 1910.119(e) The employer shall perform an initial *process hazard analysis* (hazard evaluation) on processes covered by this standard. The process hazard analysis shall be appropriate to the complexity of the process and shall identify, evaluate, and control the hazards involved in the process. Employers shall determine and document the priority order for conducting process hazard analyses based on a rationale which includes such considerations as the extent of the process hazards, number of potentially affected employees, age of the process, and operating history of the process.

> 1910.119(f) The employer shall develop and implement written operating procedures that provide clear *instructions for safely conducting activities* involved

in each covered process consistent with the process safety information.

I have highlighted some text in italics, as it pulls out three major safety requirements that must be done (again note the use of the 'shall' word).

This is not a lone example in the US of the requirement to carry out safety analysis and record the evidence – not only for a regulator or other authority, but also for the benefit of the employees in contact with the hazards. The Department of Energy (DOE) has produced a standard for the documentation of a nuclear site's hazard baseline [DOE 1995] – this might be considered to be the 'risk profile' of the site. In this standard, definition 3.15 is for a 'Safety Analysis Report', which is described as:

> Safety Analysis Report or SAR means that report which documents the adequacy of safety analysis for a nuclear facility to ensure that the facility can be constructed, operated, maintained, shut down, and decommissioned safely and in compliance with applicable laws and regulations.

Section 5.1 on hazard baseline criteria for environmental management nuclear facilities, states rather simply that "Nuclear facilities are required to develop a safety analysis report". This is a particularly good standard as it defines clearly what a SAR is and then says very clearly that you are required to have one. Excellent.

The American Petroleum Institute has also been active in this area. In 1993, API issued RP75, 'Recommended Practice for Development of a Safety and Environmental Management Program (SEMP)' for outer continental shelf oil and gas operations and facilities. Recommendations in RP75 start with an assessment of operating and design requirements and a hazard analysis. It requires establishment of safe operating procedures, working practices, and management-of-change procedures, and associated training. It calls for procedures that ensure that the design, fabrication, installation, testing, inspection, monitoring and maintenance of equipment meet safe (minimum) standards. In addition, it recommends periodic auditing of safety programs [OSMA 2002].

All these requirements, regulations, standards and recommended practices are tantamount to having a requirement for a safety case – just using different words.

Evidence for the Need to Have a Safety Case

Where not recommended or legislated for, the decision to have a safety case, safety report or safety plan often rests with the company's executives. This decision, as with all business decisions, should be justified by evidence of the need. In all companies, the most important and serious evidence comes from the financial bottom line. But in the case of 'safety' this may not be immediately obvious – you can't buy or sell 'safety', it doesn't have its own entry in the company accounts.

The value of safety may be shown in a number of ways e.g. customers moving to a 'safer' company; lost man-hours due to employee accidents; reduced operating / production hours due to increased down-time. Ultimately, this may be shown through damage to a critical piece of computer equipment or worse, the loss of a critical colleague (and I deliberately use the word 'colleague' and not the more anonymous term 'employee'), or worst of all, the death of a member of the public.

Historical Incident

On Thursday, August 11th, 1996, a 4-year-old girl called Emily was paralyzed from the chest down and her 57-year-old grandmother was killed after the miniature train ride at a fun park derailed and overturned as it approached a curve. The two victims were crushed under the weight of the rail cars . Upon investigation, the train was traveling much faster than its design speed of 12 miles per hour. The ride operator claimed to have applied the brakes as the train neared the curve, but it was discovered that many of the ride's brakes were either broken, missing, or not connected, and that most of the anti-derailment devices were missing. The speedometer was broken, along with the governor, which limits the speed of the train. The track was littered with broken ride parts. The owners of the fun park admitted negligence, but denied knowing anything about the condition of the ride prior to the accident. They have since declared bankruptcy, and most of the rides at the park were auctioned in 1997. The net result of this tragedy, besides the bankruptcy of the fun park, was a law that toughened the regulation standards for amusement park rides, appropriately named 'Emily's Law'.

Not only did this tragic incident result in the fatality and critical injury of members of the public, it led directly to the bankruptcy of the company involved, and it also caused a review and tightening of the law for the whole industry. A classic example.

Goal-based and Prescriptive Requirements

Early standards and requirements for safety performance were based on prescriptive regimes, i.e. if you complete this task you will have achieved a safe state. Prescriptive regimes work best in well established technologies with well understood failure modes. Prescriptive systems fail when there is major change, novel elements or where system complexity increases [Pitblado & Smith 2000]. The effort to maintain prescriptive requirements as capability and complexity increase rapidly becomes exponentially large, logistically impossible and out of step with technological progress. The people writing the standards are usually not necessarily those who are working with the risks, and so do not necessarily understand the problem domain best. Additionally, this had led to a culture of doing just enough to satisfy the standard, rather than actually analysing safety and reducing the hazards.

A goal-based system has been developed over the last 10 years or so to help to deal with the concern. In these types of standards, the final required level of performance is developed and defined for each problem, and this is set as the goal to be achieved. Critically, the technical means and methodology to actually satisfy the requirement is not specified. This has been left to the manufacturers and operators, often in collaboration with the regulators, to develop specific practices, tools and techniques for demonstrating goal-based safety performance.

The UK railway system has demonstrated this approach well. Between 1840 and 1870 the foundations for railway safety were conceived with a Railway Inspectorate being established, and a system for managing safety was developed based on books of rules – usually following accidents, with each accident contributing lessons to the rule books. These railway standards (known as The Blue Book) were highly detailed and prescriptive, and represented a broad encyclopaedia of collected wisdom for safe rail operations. At some point a hundred years later the complexity, cost and time of maintaining this system made it unworkable. For these reasons, and after some catastrophic accidents, the railways adopted a more goal-based safety case approach with the introduction of the Railways Safety Case Regulations in 1994 [Pitblado & Smith 2000].

When constructing a goal based safety argument, there is an implied requirement to describe evidence from both the process used to develop the system, and directly from testing and analysing the system itself.

As a analogy consider the requirement to construct a system for simply sitting on. There might be a set of prescriptive standards for height of seat, materials, number of legs, fixings to use and even texture and colours. Anything that doesn't conform to these standards will be rejected. If this were so, when I came along with my bean-bag or my inflatable chair, these would not meet many of the prescriptive standards. Yet they would be useful, comfortable and functional seating. If the standard had been goal based 'enable people to use their bottoms for support instead of their feet', any design of seat, hammock, sofa, inflatable or chaise-lounge would be acceptable.

For the example above, there would need to be a pile of research and analysis done, regarding the purpose of the seating, the profile of the sitters, how it was to be used, what might go wrong (e.g. tipping over, collapse, puncture!), the finances available, and are the woods and plastics environmentally sound? There could also be some testing done for strength, comfort and durability. In fact, all this analysis could be used as a body of evidence that provides a convincing and valid argument that the seating system was adequate for a given purpose and use profile. Perhaps we might call this a 'Seating Case'!

There are some disadvantages of having goal-based standards Generally, the purpose of a standard is to create uniformity, goal based approaches only create uniformity where there are shared competencies of practitioners and a genuine sharing of best practice. Goal-based approaches are also more difficult to assess. Non-specialists can assess prescriptive approaches, but goal-based approaches need competent assessors who can provide informed comment on correctness consistency and completeness. This requires a professional regulator or the existence of competent independent assessors.

Notes

Europa 2005, "Chemical Accidents (SEVESO) – Chemical Accident Prevention, Preparedness and Response", The European Commission, Brussels, 2005.

HSE 2005, "Charging for COMAH activities – A guide – Control of Major Accident Hazards Regulations 1999 (COMAH), Health and Safety Executive, Bootle, 2005.

MoD 2004: "Safety Management Requirements for Defence Systems Part 1" Interim Defence Standard 00:56, Issue 3. Ministry of Defence, December 2004

OSMA 2002, "Newsletter spring 2002", Offshore Marine Service Association, Louisiana 2002.

Pitblado & Smith 2000, "Safety Cases for Aviation – Lessons from Other Industries" International Symposium on Precision Approach and Automatic Landing, Munich, Germany, July 2000.

Chapter Six

Setting a Safety Boundary

What is a Safety Boundary?

The safety boundary defines all the processes, products and people that are going to be part of the safety analyses. This section will contain many of the assumptions about the system – how it operates, who uses it, where it will be used etc.

The boundary of the things that may need protection from harm extends from a person, or groups of persons, to the local buildings, nationally or even throughout the world, to the natural environment, to information and public image, and to equipment and production output. All these things may be described as being *'valuable assets'*.

It is essential to gain customer agreement on the boundary around the system, information, equipment and personnel to be included in, and excluded from, safety work. The interfaces with neighbouring systems and platforms will need to be acknowledged, as these will certainly effect your item of interest. This agreement should be recorded in the safety case report.

Specific scenarios and general ways in which the item will be operated can be used to give a richer picture for the safety assessment, again these should be agreed up front – they may even be specified for you. Any limitations of the safety assessment must be clearly stated, for example, areas where there is unclear information, limited experience or where operating conditions have not been assessed.

Historical Incident [Wears 2005]

As detailed in a case study on automation, complexity and failure from the department of emergency medicine at the University of Florida, an incident occurred causing great debate concerning system boundary. The incident was a patient needing emergency medication from an automated dispensing unit which had a number of software controlled safety interlocks. By coincidence on the day that the patient was rushed to the emergency room, the software interlocks were being re-enabled after a software upgrade process. This caused a storm of messages (one message for each medicine type) which slowed the system response to the extent of a complete stop. The human operators eventually were able to interpret the interface of the software and identify the failure mode, even though the screen message was unhelpful as "Printer not available". Runners were used to obtain the required drugs for the patient, and others were sourced from return bins next to the dispensing units.

The capture and correction of the problem did take time, it was due to human problem solving capabilities and a fortunate set of medical circumstances, that catastrophe was averted. The medical practitioners that had to work through this stressful situation were very critical of the reliability and safety of the system because the patient had nearly been lost. Hospital managers, who had the responsibility for purchasing the software system and hiring the staff, were full of praise for the system because the patient had been saved. The medical staff were considering the reliability of the system excluding themselves, and with that as the safety boundary, many would agree to its apparent unreliability and lack of safety. The managerial staff were looking at both components of the system being inside the boundary, and as such, many might also agree to its apparent safety and reliability.

It should be noted from the above example, that in neither boundary situation was it expressly said where the patient was considered to be either inside or outside of the system. You might have the opinion that for the purposes of assessing the drug dispensing system, the patient need not be considered at all. This is reasonably sensible, but how many of you would have even thought about the position of the patient, without this note being here ?

Deriving the Safety Boundary

In order for the safety case and the safety management system to exist, there has to be something or some system in place that has an amount of risk associated with it. For the safety case and safety management system to be effective, the 'thing' has to be defined. There is no prescribed methodology for deriving or defining the extent or content of a safety boundary, I am afraid it has to be figured out for each entity under consideration. However techniques such as Problem Frames can be very helpful in explicitly identifying domains and interdependencies between them to help define and communicate the safety boundary. Sometimes it may be more obvious, for example when the entity is a new product being developed, or a new assembly line or manufacturing process. However, the boundaries always throw up a number of concerns that do need to be settled early on in the safety process.

A series of questions can help – even just to get the thinking process working correctly. The first question you should ask concerns the level of the entity you are considering, is it a small component, a self-contained item, or the whole business stream? The question is this;

1. Is this thing a SYSTEM, SUB-SYSTEM or SUPER SYSTEM?

This may seem a trivial question, but the answer has ramifications to the whole safety process. The larger the entity under consideration, the more effort (and therefore resources – time, people and money) are going to be necessary. Alternatively, if the resource is fixed, then the size of the project dictates how thinly those resources are going to be spread.

The two next questions are equally fundamental to the safety analysis of the entity.

2. Where is the safety risk coming from?

3. What does the safety risk have an impact on?

Question 2 starts to focus on understanding the sources of risk within an entity, it is concerned with the problem domain. Very often the problem domain is not considered very well and the 'scientist' in all of us leaps straight to finding a solution, without spending enough time understanding the problem and the requirements for a solution. It is an all too regular occurrence for a brilliant report to be presented, only to find that it doesn't quite answer the right question.

Question 3 (still in the problem domain) begins to draw out the things, people and equipment that might be harmed by the entity. It must also be kept in mind that the answer to question 1 is very likely to be different when considering questions 2 and 3. Write down the answers and keep them close by during your work – it is likely you may have to spend a great deal of effort in the future proving and disproving these early ideas.

Boundary Diagrams

A diagram is very useful when trying to visualise a boundary of any sort, and the same is true here. Without a boundary drawn on a football field, you would not know when the ball was out of play, or when a goal had been scored. Design drawings and layout plans are useful starting points, and can also give very useful information on the mental model that the designers had when the entity was starting to evolve.

In the Safety Management System example in Chapter 3, we looked at a fire protection system in an office building. Here, I will continue with that example.

So, just to get our thoughts moving, from question 1 above, the entity we are looking at might be set as the office building. This is a SYSTEM – it is certainly bigger than a component, which might be a room in the building, and it is certainly smaller than the whole of the industrial estate or town – although there may be a discussion to be had on this point when answering question 3.

Question 2 may be answered simply for the time being, the safety risk comes from fire from within the building. Later analysis will derive where a fire would be most likely to start, but at this early boundary setting stage, fire is what we are considering as the safety risk.

It should be noted here, that the fire protection system may also be used in cases of bomb alerts, toxic release and anything else requiring evacuation. For this simple example, the fire risk will be taken as the dominant reason for the analysis.

The answer to question 3 might in this case be the same as that for question 2 – the office building. But there is the consideration of the surrounding environment – I don't mean the trees and grass around the building, I mean the next door paint factory, the hospital over the back fence, or the school in the next block. If there was a serious fire in our office building, the other buildings close by might also be effected – as indeed we might be if they had a serious fire in their buildings.

So the first go at a boundary diagram might look like this – a layout of the office building and the immediate neighbours.

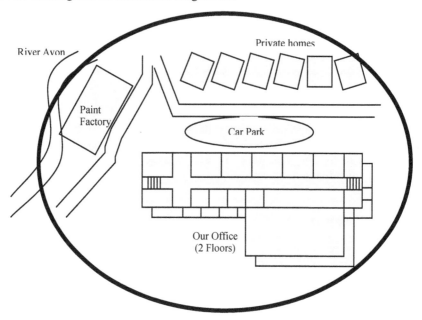

Figure 6.1 Example of a Safety Boundary

When a Diagram Might Not Work

Even in our simple example of the fire protection system in a small office, there are things that the overall safety of the system will be relying on that cannot be represented in the diagram. The diagram is focussed on geographical location and proximity layouts. In the event of our fire emergency, it is pretty much guaranteed that the fire service will be involved. They are several miles away and cannot be shown on the same local boundary diagram.

The warning is this: even when you have drawn your diagram and it looks very sensible, do not get fixated on it. You could end up ignoring the most significant factor simply because it is not in the same local or area or zone.

Other Boundary Considerations

It is vital to explicitly recognise the other factors that are within a system when considering its safety. The hardware is where everyone traditionally starts – and that is fine – for a start. The safety boundary must also consider the humans that are influenced by the system. This can be done via a list of every person involved, but this quickly becomes very inefficient. People class types are most often used. These are 'first party': the people who normally use the equipment, or who are in the building. 'second party': people who visit the building or come into contact with the equipment. 'Third party': members of the public who wouldn't come to the building or who are unaware of the equipment anyway. Even within the definition of first party there are probably different classes of user, for example, assemblers, operators, maintainers and even cleaners.

The procedures and jobs being done are another way of considering the safety boundary. Not just the normal operating conditions of production, maintenance, open and closed, but also under emergency conditions. What are the evacuation procedures? What is the job of the company safety officer?

Another dimension to consider when looking at safety is a temporal one – night and day, holiday season, lunch time. All these things affect where your personnel are and the procedures they might be doing when a potential accident initiates.

Historical Incident [NTSB 2001]

About 12:12 p.m. CST1 on January 27, 2000, a Marathon Ashland Pipe Line LLC (Marathon Ashland) 24-inch-diameter pipeline that runs 265 miles between Owensboro and Catlettsburg, Kentucky, ruptured near Winchester, Kentucky. The ruptured pipeline released about 11,644 barrels (about 489,000 gallons) of crude oil onto a golf course and into Twomile Creek. No injuries or deaths resulted from the accident. As of December 13, 2000, Marathon Ashland had spent about $7.1 million in response to the accident. During the morning of January 27, 2000, the pipeline had been shut down for previously planned maintenance work at the Catlettsburg terminal. The probable cause of the accident was fatigue cracking due to a dent in the pipe that, in combination with fluctuating pressures within the pipe, produced high local stresses in the pipe wall. Contributing to the severity of the accident was the failure of the controller and supervisors to timely recognize the rupture, shut down the pipeline, and isolate the ruptured section of the pipeline.

As noted in the incident report, the event occurred during start-up following a shut-down for planned maintenance. This was not considered to be normal operating conditions, indeed 'Start up' is considered a special case in the operation of a chemical plant. The timing of the incident was over lunch time roughly 12:00 to 14:00, the commentary on the report indicates that public reporting of oil odour and visual evidence of oil flow contributed to a shorter duration of the leak. If the event had occurred overnight or in early morning, the role of the public would not have had an effect.

The other interesting point to note is the financial cost of the accident. No injuries or deaths were reported, there was no fire, but during the following 12 months, over $7 million had been spent in response to the event. How would you have rated the severity of this incident? What descriptive phrases would you have used? These questions will be discussed in later chapters.

Notes

NTSB 2001, "National Transport Safety Board Pipeline Accident Brief PAB0102", NTSB, December 2001.

Wears 2005: "Automation, Interaction, Complexity and Failure – A Case Study", R.L. Wears MD, MS. Department of Emergency Medicine, University of Florida, Jacksonville, USA.

Chapter Seven

Measuring Safety Performance

Judging Safety Performance

In order to be able to judge safety performance and to communicate it to somebody else via the safety case or other safety report, some scale of safety measurement needs to be defined, some idea of good and bad needs to be set as a benchmark. With the wide variety of systems, procedures and materials in industry and the public domain, a scale of safety is needed to provide a measurable approach to the achievement of safety and to the exposure to risk.

Safety as a property of a system is qualitative, you could not say that a particular system was '10 safe', or that something was 50% safe. These expressions have no meaning until they are given a logical framework of reference or a measurement scale. In this way safety data collected can be subjected to a form of systematic analysis, which can turn the data into evidence directed towards judging safety performance. The data can be quantitative or qualitative, as long as the measurement system in place is geared to be able to handle either or both.

But some important questions remain. How bad is bad? How safe is safe enough? Who actually says this is bad and that is not? *The who is you.*

Public perception of risk is the key to safety. If something is thought of as being terribly dangerous, public opinion calls for action, legislation and prosecution. Our elected representatives with authority in these areas duly provide laws, working practices and procedures to be followed. They will specify things that can and cannot be done, and will apply the weight of the local, federal or national law if anything is breached. I know it's not quite as simple as that, but that is the general trend.

At the level of the individual, the perception of risk is a result of many different factors, as opposed to what might be thought of as rational judgements based on severity and likelihood of an accident. Some reasons for this departure are systemic biasing of risk information, personal experience, the use of psychological shortcuts and the way that hazard and accident information is presented to a person, including media coverage. These reasons have greater effect when the temporal separation of cause and effect is large and the immediate consequences of an action are not demonstrated by a direct link.

The relationship between public perceived risk / safety and actual risk / safety is complex and controversial. For more discussion in this area please refer to research undertaken by the UK Health and Safety Laboratory [HSL 2005].

To close this section an excerpt from this research is shown below.

> Longstanding evidence from the psychometric approach to risk perception indicates that acceptance of a hazard is related to the qualitative characteristics of that hazard. The accepted range of characteristics includes:
>
> The nature of the hazard
>   Familiarity and experience
>   Understanding of the cause-effect mechanism
>   Uncertainty
>   Voluntary exposure to the risk
>   Artificiality of the hazard
>   Violation of equity of benefits arising from hazard
>
> The risk consequences
>   Ubiquity of the consequences
>   Fear of the consequences
>   Delay effect
>   Reversibility
>   Social and cultural values
>
> Management of the risk
>   Personal control over the risk
>   Trust and distrust in the perceived institutional control of the risk
>
> There is an indication that hazards should be judged on a case-by-case basis to account for the separate contexts (characteristics) of each hazard.

NOTE: I would just like to refer the authors of the above text to the discussion on the derivation and meaning of hazard and risk – see chapter 2.

Measurement Scales

To communicate about risk, there is a need for some benchmarks or positions on a scale to have an recognised comparison. For a scale to be useful it needs to have a particular set of characteristics. The scale needs to be exhaustive, such that it can represent any possible outcome; and it must be exclusive, such that any possible outcome can only be represented by one point on the scale. Scales are usually one of five types according to the need of the display – Nominal, Ordinal, Serial, Interval and Ratio.

A nominal scale meets the two basic criteria for a scale. It is a simple listing of the complete set of all possible alternatives and the variables associated with them, for example, the names of students in a class. The scale has all the students on it and each student only fits into one point on the scale. The points on the scale are discrete and do not imply anything about the completeness of the scale, or any intermediate points.

The ordinal scale is similar to the nominal scale, but the items imply some order. For example, days of the week or months of the year. The frame of reference has a finite boundary, there cannot be a 13th month. Again, all items are listed and each item has one point on the scale.

The serial scale is like the ordinal scale, but with the points on the scale having a regular and discrete characteristic. For example, the calendar has a space for each year (even if no data items occurred during that year), and you can only be in one year at a time. The frame of reference has no end points to the scale, but you just can't have the point 2006½.

The interval scale is like the serial scale, but the values on the scale are not discrete, they are continuous and infinite. Not only can you have 0.5, you can have 0.5555555, you can have any value you like.

The ratio scale is a further step on from the interval scale, except that it has an absolute zero as a starting point. This means that every value is absolutely comparable to every other. For example, world population or hours of sunlight, it is also a requirement that the scale cannot have negative values.

Safety Measurement Scales

From the definitions discussed in this book, safety and risk are related, safety may even be treated as a freedom from risk. Risk, and therefore safety also, are considered to be composed of two features – the severity of an event and the frequency of an event. Each of these need some scale to judge the parameter so that a combination of the two may lead to a scale type and measurement regime for risk and safety.

Initially it does appear that any of the previous scale types might be possible. Indeed, the nature of frequency and severity seems to point towards a complex ratio scale – there being an easy absolute zero in both cases. But this actually appears to have fallen away, or not yet been adopted through being too difficult to interpret. Others appear not to be used through being too imprecise to be of value (nominal scale). Ordinal scales are certainly used, and these reflect qualitative safety scales; quantitative safety scales have been defined in both ordinal and serial terms.

Event Severity Scales

When you are considering the safety of an item or a system, there are events that can happen that result in bad effects. To be able to communicate about the levels of impact of these events, a scale is defined and used. Initially this can start with two points on the scale 'good' and 'bad'; this can be extended to have 'very good' and 'very bad'. However, this is where trouble starts to creep in, does 'good' have a real meaning when discussing severity? How bad is 'bad', and how much worse is 'very bad'? Could we perhaps introduce 'very, very bad'? This simple concept is quickly shown to be not sufficient without further detailed definition of the terms.

Another concept used is to just have a series of numbered categories, say '1' to '6'. Where '6' has the worst severity and '1' has the least severity. The European scale for industrial accidents [BARPI 2005] uses this methodology on

18 different parameters relating to accidents – for example, the number of people killed, the amount of substance released, the cost of clearing up and the environmental consequences. For the numbers of persons killed in an accident (H3 scale denoting human aspects – 3rd scale) the 6-category scale is as follows:

- None
- 1 employee
- 2-5 employees and/or 1 external rescue person
- 6-19 employees and/or 2-5 external rescue persons and/or 1 public
- 20-49 employees and/or 6-19 rescue persons and/or 2-5 public
- Over 49 employees and/or over 19 rescue and/or over 6 public.

There are difficulties with this sort of scale, especially when comparing it with the 6-point scale for environmental consequence – how many human lives are equivalent to a bird sanctuary? But the main problem here, I believe, is that the scale type is confusing. The use of a numerical scale implies that we have a serial, interval or ratio scale type. It can imply that there is an equal spacing between the category steps, or that a level '4' accident is considered twice as bad as a level '2' accident, which is clearly not the case. This scale is actually an ordinal scale type, which has 'None' and 'Over 49' as the end points, and what appears to be an arbitrary arrangement of the steps for each category.

Further scales (H4 and H5) categorise the number of persons with injuries needing hospitalisation and those with just minor injuries. Again though, the comparison between a level '5' minor injuries category (200 to 999 employees and rescuers and/or 20 to 48 members of the public) and a level '5' death category (as above) is difficult to rationalise.

In the US Department of Defense (DoD), accidents are classified according to the severity of resulting injury, occupational illness, or property damage. Property damage severity is generally expressed in terms of cost and is calculated as the sum of the costs associated with military property and non-military property that is damaged in a DoD accident. Additionally, if injury or occupational illness results, an event is reportable even if the associated costs are less than the minimum dollar criteria. As previously noted, the use of the term 'accident' has particular legal implications under US law, so the term 'mishap' is often used instead.

> Class A Accident. The resulting total cost of damages to Government and other property is an amount of $1 million or more; a DoD aircraft is destroyed; or an injury and/or occupational illness results in a fatality or permanent total disability.

> Class B Accident. The resulting total cost of damage is $200,000 or more, but less than $1 million. An injury and/or occupational illness results in permanent partial disability [which does have its own definition]. Or when three or more personnel are hospitalized for inpatient care (which, for accident reporting purposes only, does not include for just observation and/or diagnostic care) as a result of a single accident.

Class C Accident.   The resulting total cost of property damage is $20,000 or more, but less than $200,000; a nonfatal injury that causes any loss of time from work beyond the day or shift on which it occurred; or a nonfatal occupational illness or disability that causes loss of time from work or disability at any time (lost time case).

Class D Accident. The resulting total cost of property damage is $2,000 or more but less than $20,000.

Class E Accident. The resulting damage cost and injury severity do not meet the criteria for a Class A-D accident ($2,000 or more damage; lost time/restricted activity case). [DoD 2000].

In this classification, the scale type does not rely on numerical labels, and is quite clearly ordinal in nature. The use of the alphabet gives a frame of reference, in that 'A' is at the top and there is an order below, although it is impossible to say whether 'A' is twice as bad as 'C'.

One concern with these types of description is that the monetary terms used in the scale are fixed, and in real terms the value of property, equipment and life increases. So over a short period of time, more accidents become classed as 'A'.

A further classification scheme in use in a variety of industry types the UK uses a nominal scale label, but an ordinal description. It appears to work fairly well, for example from the UK defence industry [DOSG 2003]:

Cataclysm:	Thousands of fatalities and injuries
Disaster:	Hundreds of fatalities and injuries
Calamity:	Tens of fatalities
Catastrophic:	Single figure, multiple deaths (up to 10)
Critical:	Single death and/or multiple severe injuries
Severe:	Single or several serious injuries
Significant:	Several minor injuries – significant lost time
Minor:	Single minor injury.

Like the US DoD version, a monetary value can be assigned to each category, and further definitions may be added for environmental, political and equipment domains.   This type of classification scheme does start to take on the characteristics of a ratio scale. There is a pseudo absolute zero defined (below a single minor injury) and there is an apparent orderly increase in each of the categories. This does mean that there is an explicit consistent ratio between the categories, unlike the European example scale, which can be difficult to interpret. This also combines together the death and injury impacts, giving a relationship between them of 1 death $\approx$ 10 severe injuries. This ratio is certainly still open for ethical debate.

One of my personal favourite safety scales is concerned with the magnitude of media coverage as the variable. This can really be useful when trying to focus corporate minds on safety, when a product or system that they have produced could have this sort of negative publicity impact [ibid.]:

- Cataclysm: World media outcry
- Disaster: International TV headline news
- Calamity: International TV news over several days
- Catastrophic: International news coverage, national TV news over
                         several days
- Critical: National TV news coverage, local news coverage over
                         several days
- Severe: Considerable local coverage with inside notes in national
                         press
- Significant: Noted in local press
- Minor: No outside interest.

Event Frequency Scales

To start this section, I want you to start to think about the answer to this question;

"How often do accidents happen?"

Well, you should immediately ask me what sort of accident am I talking about? How severe an event is being considered? ... there is probably a whole raft of answers you would need to know before being able to consider the answer to that first question.

In a similar way to severity, to be able to communicate about a bad and not-so-bad frequency of events a scale has to be defined for the system of interest. The most obvious one that is used is the human calendar – hours, days, weeks, months, years. This does appear to be a most convenient scale, ready made for use in discussing the occurrence of anything. Most events we hear about in the media are quoted in units of time, most variables we come across in our human day-to-day lives are concerned with time. For example, miles per hour; hours of T.V. per day; burgers per week, monthly income, spend per quarter, crimes per year. It is very easy to link these units with the occurrence of accidents. However, accidents do not always occur in relation to time.

Consider communicating about the frequency of motor-vehicle accidents. Does it matter how much time you spend in a car, or how far you drive? For a single journey, these two variables are related by the speed at which you go, so does this mean that travelling faster (less time in the car/truck) is actually safer? Please submit your answers in a technical paper to the next automotive safety conference!

There are multiple factors involved in the frequency of road traffic accidents; time of day, age of driver, condition of the road surface, weather, class of road user, etc. In the UK, road traffic accidents do have a consistent severity definition for data comparison purposes, this has been chosen because it is the law to report such incidents, so the data set is particularly rich. The term used is 'KSI' which means 'killed or seriously injured'.

The UK Department for Transport has the following definition for a serious injury on the roads:

Serious injury: An injury for which a person is detained in hospital as an 'in-patient', or any of the following injuries whether or not they are detained in hospital: fractures, concussion, internal injuries, crushings, burns (excluding friction burns), severe cuts and lacerations, severe general shock requiring medical treatment and injuries causing death 30 or more days after the accident. An injured casualty is recorded as seriously or slightly injured by the police on the basis of information available within a short time of the accident. This generally will not reflect the results of a medical examination, but may be influenced according to whether the casualty is hospitalised or not. Hospitalisation procedures will vary regionally.

The recording of KSI accidents is done over three different units – kilometres travelled, number of journeys and number of hours. A comparison of the main transport modes is shown below [DfT 1999] to indicate how the units of frequency of accidents can have very different results on the way the risk of the transport modes are viewed.

Table 7.1 Comparison of Transport Accident Frequency Units

KSI per passenger km *(100 million passengers)*		*KSI per passenger journey* *(100 million passengers)*		*KSI per passenger hour* *(100 million passengers)*	
Air	0.007	Bus/coach	14	Air	7
Bus/coach	2	Van (truck)	23	Bus/coach	35
Van (truck)	2	Air	28	Van (truck)	84
Car (auto)	4	Foot	50	Water	90
Water/boat	5	Car (auto)	55	Car (auto)	171
Foot	63	Water/boat	168	Foot	257
Pedal cycle	88	Pedal cycle	328	Pedal cycle	1073
TWMV	139	TWMV	1623	TWMV	5382

TWMV stands for 'two-wheeled-motor-vehicles', so when your parents said that motorbikes were dangerous, they were right. Rail was not included in the data set because of a change in the definition of 'rail travel' part way through the data collection time.

Aside from the interest in the data values themselves, the discussion is only possible by having a useful set of frequency units. In this case three different unit sets are used to give an overall picture of performance. In other industries, specific unit types will be more preferable, and again it is most likely that multiple sets of units will be used to give the richer information.

In the process industry, the units may be per operating hour and/or per volume produced. In the building industry, the units may be per hour and/or per worker year, in the military, the units may be per firing, per flying hour and/or per mission.

Once the units of frequency have been established, the separate classes along a scale in those units need to be defined. As with severity, there are a number of types of scale in use. The nominal scale label and ordinal description is popular, with a typical example being something like this:

Frequent	Likely to be continually experienced
Probable	Likely to occur often
Occasional	Likely to occur several times
Remote	Likely to occur some time
Improbable	Unlikely, but may exceptionally occur
Incredible	Extremely unlikely.

Without the explanatory notes (and some might say even *with* the notes), it can be difficult to judge the correct order of these phrases. From your interpretation of the words, does 'probable' really mean something more frequent than 'occasional'? Even once the order has been accepted, the interpretation of the meaning for a specific product or system is still something that takes many people many hours.

The DoD probability level description gives more detail [DoD 2000];

Frequent (A)	Likely to occur often in the life of an item, with a probability of occurrence greater than 10-1 in that life. Continuously experienced.
Probable (B)	Will occur several times in the life of an item, with a probability of occurrence less than 10-1 but greater than 10-2 in that life. Will occur frequently.
Occasional (C)	Likely to occur some time in the life of an item, with a probability of occurrence less than 10-2 but greater than 10-3 in that life. Will occur several times.
Remote (D)	Unlikely but possible to occur in the life of an item, with a probability of occurrence less than 10-3 but greater than 10-6 in that life. Unlikely but can reasonably be expected to occur.
Improbable (E)	So unlikely, it can be assumed occurrence may not be experienced, with a probability of occurrence less than 10-6 in that life. Unlikely to occur, but possible.

The additional notes in the DoD standard state that mishap probability can be described in terms of potential occurrences per unit of time, events, population, items or activity, as appropriate. But there is no further guidance. As we have seen, the relative safety appearance and ranking of events can be dramatically altered if you modify the units used.

The Risk Matrix for Communicating About Safety

The severity and probability characteristics introduced in this chapter can be combined in multiple ways to describe a single event. One combination might be a class 'A' severity paired with a class 'D' probability; or a 'Critical' severity paired with a 'Remote' probability. Of course you could get 'n' multiplied by 'm' combinations, where 'n' is the number of severity classes and 'm' is the number of probability classes.

Each combination will have a safety or risk implication. If the combination is a high severity and a high probability, this will have a higher risk rating that a combination of a low severity at a low probability. The two factors may be displayed in a matrix table as shown in Table 7.2;

Table 7.2 Display of Probability and Impact in a Combined Matrix

	Calamity	Catastrophic	Critical	Major	Minor
Frequent					
Probable					
Occasional					
Remote					
Improbable					

The matrix may be populated with labels of risk priority as shown in Table 7.3;

Table 7.3 Matrix Populated with Risk Priority Classes

	Calamity	Catastrophic	Critical	Major	Minor
Frequent	A	A	B	C	C
Probable	A	B	B	C	D
Occasional	B	C	C	D	D
Remote	C	D	D	D	D
Improbable	D	D	D	D	D

They may even be coloured with red for 'A' and blue for 'D', with shades of yellow and green in between. Of course the meaning of the 'A', 'B', 'C' and 'D' needs to be made. Again, as you might expect there are a few different interpretations and implications, with different objectives depending on what you are going to do when you know about the risk level. This does link back to the earlier chapters on the purpose of a safety case.

One typical description might be as follows:

'A'   An intolerable risk, which can only be accepted under extreme circumstances.
'B'   A tolerable risk, providing the benefit received grossly outweighs the risk.
'C'   A tolerable risk, providing the benefit received outweighs the risk.
'D'   A broadly acceptable risk, which may be treated as negligible.

Each project or system may set what actions to take if the risk is in a particular class, for example:

'A'   Operation cannot be accepted, limitation in use must be declared.
'B'   Operation can be accepted with evidence of gross benefit received.
'C'   Operation can be accepted with evidence of benefit received.
'D'   Operation can be fully accepted with evidence of low risk.

Additionally, the project may set the authority level required in order to accept a risk in any of the risk classes, for example:

'A'   Only acceptable with approval from government minister.
'B'   Only acceptable with approval from board of corporation.
'C'   Acceptable with approval of project leader
'D'   Acceptable with approval of project safety engineer

Obviously, no one wants to have to put limitations on their system or product, as this would be uneconomic and not at all popular with the customers, accountants or investors. It would be much better all round to work in the 'B', 'C' and 'D' categories, but here there is a burden of providing evidence, i.e. you have to do some work. This will be covered in future chapters, but first there is more discussion to have concerning safety and risk scales.

Interlude: A Note about Examples BecomingFfact

This chapter has quoted from several international standards on severity and frequency class concepts. It has gone on to discuss the combinations of these two factors to enable rating and communication about risk and safety. Two statements need to be made about them;

THEY ARE ONLY THE OPINIONS OF AUTHORS
THEY ARE ONLY EXAMPLES.

A particular example definition of catastrophic of 20 to 49 civilian fatalities is only that − it's an example. In some specific industry sectors, the scales of severity, probability and risk will be mandated. In many others, there is no set standard other than the undertaking of safety analysis. The matrix formation is one methodology to follow, but that means you have to define your own matrix appropriate to your own project. Copying mine or anyone else's without good reason will leave you open to criticism, these examples are just that - examples.

Populating a Risk Matrix

Consider again our risk matrix that was developed earlier in this chapter (Table 7.3 is shown again):

Table 7.3 Matrix Populated with Risk Priority Classes

	Calamity	Catastrophic	Critical	Major	Minor
Frequent	A	A	B	C	C
Probable	A	B	B	C	D
Occasional	B	C	C	D	D
Remote	C	D	D	D	D
Improbable	D	D	D	D	D

How did it come to be laid out in this fashion? Well, it is from knowledge of the meanings of the axis labels and risk classes, the functions of the system, the benefits it produces and the perception of risk associated with the system.

So even before we get to carrying out any risk and safety analysis for our safety case, we need to set all the ground definitions and fix up the way that safety is going to be judged – the whole point behind this chapter, hopefully now demonstrated.

A useful starting point for populating a risk matrix is the fixing of the points where a single fatality is intolerable and where one is considered broadly acceptable. Let's start with a blank matrix, we will develop definitions for the axes as we go through.

The description of the severity definitions is the column to look at first. For this purpose, we will say that Catastrophic defines a single fatality – it may also define multiple injuries, environmental damage levels, media coverage, but let's start with a single fatality – most people can identify with that. The frequency classes might be defined as follows:

Frequent	Weekly
Probable	Monthly
Occasional	Annually
Remote	10 Yearly
Improbable	100 Yearly

Let us suppose that we are considering European rock climbing as an activity. With the severity set at a single fatality, what perception of intolerability do you have for the frequency of a single fatality? You might reason that if people were being killed on a weekly basis across Europe, the sport would soon disappear, or it would have to be severely restricted and controlled. If monthly works out as 12 fatalities per year, well that might be acceptable. A single fatality once per year from rock climbing over the whole of Europe might be seen as being excellent. The discussion goes on, perhaps including the opinion and consensus of a team of the relevant people involved – climbers, mountain rescue, rope manufacturers, an instructor etc.

For the sake of our example, weekly fatalities are unacceptable and monthly fatalities might just be tolerated. We now have the first two points to be displayed in our risk matrix, see table 7.4.

Table 7.4 Risk Matrix Showing Intolerability of Single Fatality

	Calamity	Catastrophic	Critical	Major	Minor
Frequent		A			
Probable		B			
Occasional					
Remote					
Improbable					

The next step is to work up this column from the bottom and fix the point where the negligible point would come in. If there were no rock climbing fatalities over a 100-year period in the nation, this would probably be considered very acceptable. This point would be a 'D'. For a single fatality over a 10-year period, again probably an acceptable frequency, = 'D'. For a single fatality per year, well you might not accept that as a wanted situation, but it might be tolerated. Therefore not a 'D', so a 'C' instead.

So we have now populated our single fatality column completely.

Table 7.5 Risk Matrix Showing Risk Classes for Catastrophic Impact

	Calamity	Catastrophic	Critical	Major	Minor
Frequent		A			
Probable		B			
Occasional		C			
Remote		D			
Improbable		D			

There are now two ways to progress the completion of the table. Firstly, the filling in task can then be done for each severity column, and cross-checked horizontally through each probability row. Alternatively a series of useful, loose rules can be employed, and this can be done globally over the whole matrix, or locally to portions of the matrix.

For example, if 'catastrophic' was unacceptable at 'frequent', then it's a certainty that 'calamity' is also unacceptable. In a similar vein, if 'catastrophic' was broadly acceptable at 'remote' and 'improbable', then the lower severity categories are also going to be considered the same. So, the matrix now looks as shown in Table 7.6;

Table 7.6 Risk Matrix Showing Initial 'A' and 'D' Risk Classes

	Calamity	Catastrophic	Critical	Major	Minor
Frequent	A	A			
Probable		B			
Occasional		C			
Remote		D	D	D	D
Improbable		D	D	D	D

The final assumption (which MUST be declared if you are employing it), is that each higher severity and frequency category is roughly one order (i.e. x10) greater than the one preceding it, AND that the risk class interpretations are also roughly one order apart.

This means that whilst 'catastrophic' at 'frequent' is considered 'A', a 'critical' at the same probability is considered one risk class lower, i.e. 'B'. Likewise a 'major' at the same probability is considered one risk class lower, i.e. 'C'. So the matrix starts to become more populated, see Table 7.7;

Table 7.7 Risk Matrix Showing 'Frequent' Probability Category

	Calamity	Catastrophic	Critical	Major	Minor
Frequent	A	A	B	C	D
Probable		B			
Occasional		C			
Remote		D	D	D	D
Improbable		D	D	D	D

This rough rule can then be employed for all the remaining points in the matrix, as shown in Table 7.8;

Table 7.8 Completed Risk Matrix

	Calamity	Catastrophic	Critical	Major	Minor
Frequent	A	A	B	C	D
Probable	A	B	C	D	D
Occasional	B	C	D	D	D
Remote	C	D	D	D	D
Improbable	D	D	D	D	D

The starting point in this example was the single fatality event, the position of what was intolerable was determined through scientific reasoning and expert perception and opinion. This is a perfectly acceptable method for developing a risk matrix. It does have to be noted however, that in some cases the starting point may be already specified. This might be from legislation, codes of practice or simply in the contract you are trying to meet. Example starting points might be, the system will not have a risk of fatality greater than 1e-3 per worker day; it is unacceptable for this product to cause injury at a rate greater than 1 per 1000 operations. Again, the units are very important and the expression of how these statements are made has to be mutually understood between publisher and reader. If in doubt ask, these statements are likely to become your safety targets, upon which you and your system or equipment will be judged.

Special Note

As ever, the previous derivation is just my example, employed to show how risk analysis MIGHT be initiated. Please don't copy it, refer to it by all means, but do not assume that my example will be suitable or relevant to your system, product or software, it probably will not. I will not accept any liability for you not being bothered to think about and generate your own risk assessment matrix.

The Layout of a Risk Matrix

You will note that the risk matrix developed here has the highest values of probability and impact to the top-right of the matrix. There is no real convention that says this must be so. It is tradition to have the origin at bottom-left, with the axes increasing up the page (y) and to the left of the page (x). In these examples, the probability axis does follow tradition, but it may be argued that the impact axis starts large and gets smaller, so should be placed the other way around. That is an acceptable orientation, but the impact definitions in the examples here are all negative, and so are placed to the left of the 'x' origin, and getting larger to the left. The layout is correct.

You will probably see matrices laid out in several different ways – as long as you can understand them, they are all perfectly fine.

The Final Check

When the risk matrix is completed for the first time, it does need a sanity check for gross errors. Do the risk classes actually look correct, or is something not sitting quite right? This is an important check and it is usually at the limits that the potential problems occur. For example you may find that you have defined the 'improbable' and 'calamity' point as a 'C'. This may be correct, but it does mean that absolutely every event that could credibly be a calamity, even if it is less probable than 'improbable' (which should contain all lower probabilities due to the scale type being used here), the event will still need to have all the 'C' type actions carried out on it.

Also, the point 'frequent' 'minor' is a key location. If this is not a 'D', all the most minor occurrences at a 'frequent' probability will have to have the 'C' type actions. This may be onerous to the project and use a lot of resource.

At these two locations, sometimes a formal, rigid risk matrix is too restraining and needs to be allowed to bend.

The other crucial point that does cause a lot of debate is the 2-2 or 3-3 point on the leading diagonal. In the example, the 2-2 point is at 'catastrophic' 'probable'. The argument goes that the sanity check can highlight that this type of point really needs to be one risk class higher. As far as any specific matrix goes, it doesn't really matter what risk class this point (or any of the others) actually is, as long as the reason is recorded within the safety case having been accepted by the customer or regulator.

Summary

Communicating about risk and safety needs a common understanding about the properties of risk and safety. These are commonly accepted as severity and probability, and these combine together to produce a description of risk. This can be formalised into a matrix formation, which can be used to prioritise risk and so can lead to the development of an action list to reduce risk and improve safety.

Notes

BARPI 2005, "European Scale for Industrial Accidents", Bureau d'Analyse des Risques et des Pollutions Industrielles (BARPI), Lyon, France, 2005.

DfT 1999, "Transport Safety", Consultation Paper Table A1.3, Department for Transport, London, 1999.

DoD 2000, "Standard Practice for System Safety" MIL-STD-882D, US Department of Defense, 10 February 2000

DOSG 2003, "DOSGST2 Risk Matrix Definitions Table", Defence Ordinance Safety Group, MoD, Abbeywood, UK, 2003.

HSL 2005, "Review of the Public Perception of Risk, and Stakeholder Engagement", HSL/2005/16, Health and Safety Laboratory, Buxton, UK, 2005.

Chapter Eight

Safety Targets

The Role of Safety Targets

The whole concept of risk and safety analysis is predicated on the notion that there is some level of risk exposure that is acceptable. I do not intend to contradict this. However, there are difficulties with having a fixed line in the sand that mandates that one side is safe and the other side is not. This is very difficult to justify and control.

Consider for a moment the concept of road traffic speed limits. Irrespective of what they actually are in your country and the units of speed used, what is the purpose of them, what is the implication of driving above them? Are speed limits the marker between safe and unsafe driving speeds? That certainly appears to be a correct interpretation. But as we have seen in the previous chapter, having a universal speed limit for road users has not equalised the accident rates for each user type. If you do believe that speed is a/the major contributor to road traffic accidents, it might be prudent to have different speed limits for each road user type, with three wheeled motor vehicles having the lowest limit. Indeed this sort of thing is already happening with large heavy trucks – the larger and heavier the truck is, the lower the speed limit is. This is not necessarily declared on the roads and highways, but on the rear of the vehicle in question.

These limits and other 'safety targets' are fixed points along the scale of risk by which people can assess performance. In a few examples there are legislative implications for exceeding a safety target, but may be better described as safety limits – as in the fines that are administered for (being caught) speeding.

So it is essential to realise that safety targets are not safety limits. Targets may be exceeded, but there are implications for doing so - extra regulation, additional impact reduction measures, limits on the operational capability. In the UK, it is not uncommon for pedal cyclists to be cautioned and fined for travelling too fast. When using motorised two-wheeled vehicles on the roads in the UK, it is mandatory to wear a crash-helmet, this is not the case for un-powered cycles, even though on down-hill sections, near equivalent speeds can be reached – so I'm told of course!

Setting a Safety Target

There are several ways of setting safety targets, and for each safety case the scale used and the specific points of good and bad need to be defined. These fundamentals are absolutely essential.

Quantitative Targets

Quantitative targets are usually established, they give an obvious indication of a position on a scale for anything including safety. The UK Health and Safety Executive has carried out research in the area of setting quantitative safety targets. Guidelines from the HSE suggest that no one person should carry an unreasonable amount of risk. A tolerable limit for a fatality risk is suggested for an individual at a quantitative value of a probability of 1 in 1000 per worker-year for all working tasks (including, but not exclusively, the use of hazardous materials and equipment). This means that if the calculated probability of an individual's fatality from all working tasks has the predicted occurrence of greater than 0.001 per year, then that person's exposure to risk is deemed to be intolerable.

At the other end of the safety scale, there is a level of risk exposure that is considered broadly acceptable. Strange as it may seem, there really is some level of risk of fatality that is considered to be OK. The question is, what is that level? ... And perhaps more importantly, how is that level justified? Not only, how is it justified to some authority that is going to inspect the safety case, but also, how is it justified to the person who is actually going to be exposed to that risk.

This view of safety may be represented graphically showing the limit of tolerable risk exposure and the level below which is considered to be broadly acceptable. These boundaries can be annotated with quantitative (or qualitative) indications, such that a system's safety performance may be judged.

Between these two levels, there is the region of risk tolerability. In the region, the risk of using the equipment, system or software is deemed to be acceptable if it can be demonstrated that the cost (in terms of time, trouble and expense) of reducing the risk further would be 'disproportionate' to any improvement gained [Engineering council 1993]. An alternative interpretation is that within this region, the risk is only taken if a benefit is desired [ibid.].

This concept has also become known as the 'as low as reasonably practicable' (ALARP) principle. This will be discussed in more detail in the next chapter.

Target Apportionment

A series of consistent safety targets may be derived for all contributory components to the system in question. For example, the safety performance of an aircraft is contributed to by multiple key component performances – structural integrity, fuel supply, engine performance, hydraulic system and the pilot-cockpit interaction. As each of these (and more) systems contribute to meeting the safety target for the whole aircraft, it is entirely appropriate that each system should have its own proportion of the overall safety target. As such, it is common for the safety target of a system to be apportioned out over the component parts.

In some instances the overall target is simply divided equally between all components. This is easy to implement and easy to understand, but it can lead to difficult interpretations of safety budgets. Where this difficulty does happen, trading of safety budgets starts to take place. Again, this is perfectly acceptable, if it is done for the right risk-based reasons and recorded.

An alternative method is to base the proportion on the risk associated with each component. In the simple aircraft example above, it might be argued that the pilot-cockpit interaction has the highest risk, as this is the fundamental system for the pilot being able to fly the craft. As such, it should be able to hold a higher budget of the overall safety target.

The key part of the last paragraph is 'it might be argued' The arguing part can only be substantiated by a full understanding of the risk contribution of the components of the system. This full understanding can only be achieved through systematic hazard and risk analysis, which of course should be recorded as high quality evidence for the safety case. A simple taxonomy may be set for this process as follows;

- Establish the whole system safety target (from legislation, standards or prior use)
- Derive a risk measurement method (see Chapter 6)
- Risk assess the system components (likelihood and severity)
- Prioritise the system components according to risk profile
- Apportion the whole target over the components according to priority.

Whatever the method chosen, a series of consistent targets should be derived for all individuals, groups, equipment types and other valuable assets that are the subjects of the safety case argument. The exact method and rationale is up to the safety team involved, but they must be recorded in the safety case, so that future designers, users and regulators can understand why targets were set in a particular way.

Quantitative Targets in Use

As part of its 2004 work programme, the Civil Aviation Authority of New Zealand began to develop aviation safety targets to be achieved by the year 2010. In September 2004 a consultation document was released, which discussed quantitative safety targets in a very clear and useful way [CAA NZ 2004]. The CAA recognised that while it would be highly desirable for aviation activities to be completely safe, it is generally recognised that this does involve some level of risk, and that when there is risk, it is not possible to have absolute safety. Their key question was 'How safe should New Zealand aviation be?' The existing safety targets were set quantitatively in the form of accidents per 100,000 hours of normal operating activity, and were published as a goal for the aviation community to aspire to.

The key benefits that the CAA (NZ) had identified in setting safety targets are that;

The targets will provide a strategic goal for the CAA and provide links to its corporate vision of 'New Zealand aviation – free from safety failure'.

The targets will implicitly provide a means for monitoring the success of the aviation community as a whole in influencing safety outcomes.

Safety targets enable Government, the public and the aviation community to measure safety performance against appropriate yardsticks.

The new targets set for New Zealand aviation have been set in the quantitative unit of the NZ$ worth of social cost. The safety target structure has been split 13 ways to include public air transport (5 sub groups), commercial use (5 sub-groups) and non-commercial use (3 sub-groups). Each of these has its own social cost target to be achieved by 2010.

For example, target group 1 is for Airline operations of large aeroplanes, including training, test, passenger and freight, domestic and international. It includes aeroplanes that have a seating capacity of 30 seats or more, and/or a payload capacity of more than 3410Kg. The target has been expressed in accident cost per seat hour, and for this class of aviation the current estimated rate is NZ$0.13, and the target for 2010 is to be less than NZ$0.10 [CAA NZ 2005]. For more information, please see the reference.

The advantage of expressing targets like this is that they start to reflect public perception and the way society feels about accidents and injuries. Also because they have been worked out in monetary terms, some direct comparisons can be easier. The disadvantage is that safety targets expressed this way are not in common practice, particularly in the aviation domain, so some direct comparisons are much harder. It is also pretty hard for the general public to understand whether $0.10 is just good or fabulous – the reference scale is difficult to apply – see chapter 7.

Qualitative Targets

Qualitative targets can be used to supplement quantitative targets, or they can be used on their own when a quantitative target cannot be applied, or when one would not make logical sense, but they may still be based on classifications of combining severity and probability of potential accidents. For example, a catastrophic accident having no more than a most unlikely probability of occurring. Of course, the words 'catastrophic' and 'most unlikely' still need those careful definitions.

Qualitative targets, and evidence to measure against the targets, need to have special and focussed effort if they are to be used satisfactorily. Qualitative details are, by their nature, impossible to prove absolutely. Qualitative evidence falls into one of two categories – it is evidence *of* some claim or evidence *for* some claim. The distinction is crucial, evidence *of* only provides a possible condition; evidence *for* provides a necessary condition and as such, is treated as stronger evidence for interpretation [Miller & Fredricks 2003].

Simply because excessive speed is involved in many road traffic accidents, doesn't necessarily mean that excessive speed is the cause of the accident.

In safety cases, qualitative targets and evidence can be used in retrospective cases where quantitative data was not recorded. Accident records may be available, but hours running, distance travelled, age or size has not been noted down. So a retrospective target may have been 'no catastrophic accidents' for some piece of equipment. Records and anecdotal evidence will provide a source of details about historical incidents, which can be qualitatively assessed to see if there had indeed been any 'catastrophic' accidents.

In safety cases that are concerning contemporary equipment or system, qualitative targets and evidence can be used, if they are controlled carefully, and there is a clear and accepted (and used!) meaning of the critical definitions. These can be defined mathematically (but then why not stick to quantitative targets?), or syntactically, using a commonly agreed safety language. Very often, I have seen the phrase 'qualitative' used where the quantitative target is just written in words and not numbers, this is obviously nonsense.

Notes

CAA NZ [2004] "Safety Consultation Document (September 2004) – Setting Aviation Safety Outcome Targets for the Year 2010", Civil Aviation Authority of New Zealand, Lower Hutt, 2004.

CAA NZ [2005], "New Zealand Aviation Safety Outcome Targets to be Achieved by 2010", Civil Aviation Authority of New Zealand, Lower Hutt, 2005.

Engineering Council [1993], "Guidelines on Risk Issues", The Engineering Council, London, 1993.

Miller & Fredricks [2003], "The Nature of Evidence in Qualitative Research Methods", International Journal of Qualitative Methods 2 (1) Winter, 2003.

Chapter Nine

So Far as is Reasonably Practicable

So Far as is Reasonably Practicable

The phrase 'so far as is reasonably practicable' is a critical wording used in UK law on health and safety. It arose through a court trial in 1949 where the National Coal Board was in court defending against the death of an employee, Mr Edwards.

Historical Incident

Mr Edwards was killed when an unsupported section of a travelling road in a coal mine gave way. Only about half the whole length of the road was actually supported properly. The coal board argued that the cost of shoring up all roads in every mine in the country was prohibitive and not reasonable when compared to the risk. The legal question at issue was not the cost of shoring up all roads in every mine operated by the company. The issue was the costs of making safe the specific sections of road that were at risk of falling in. Some, perhaps many of the company's roads were secure and showed no signs of failing – they didn't need extra support. Others had already have fallen and had already been repaired. The section in question was already supported by timber along half its length. The cost of making it safe was not great compared to the risk of injury and loss of life.

The test for what is reasonably practicable was set out in this case. The case established that the risk must be balanced against the 'sacrifice', whether in money, time or trouble, needed to avert or mitigate the risk. By carrying out this exercise the employer can determine what measures are reasonable to take. If the measures to be taken are grossly disproportionate to the risk, then the measures need not be taken.

The full legal definition was set out by the Court of Appeal (in its judgement in Edwards v. National Coal Board, [1949] 1 All ER 743), and is as follows:

> Reasonably practicable is a narrower term than 'physically possible' ... a computation must be made by the owner in which the quantum of risk is placed on one scale and the sacrifice involved in the measures necessary for averting the risk (whether in money, time or trouble) is placed in the other, and that, if it be shown that there is a gross disproportion between them – the risk being insignificant in relation to the sacrifice – the defendants discharge the onus on them.

The case effectively gave an implied requirement for carrying out risk assessment and recording it somewhere. Although a specific call for a safety case was not made, the implied requirement of risk assessment and recording evidence pretty much covers the role of a safety case.

The legal background to the SFAIRP definition is actually not risk based at all – it is sacrifice based. The concept is to adopt measures to avert the risk, except where they are ruled out by being grossly disproportionate to the risk. That is to say, look at all the physically possible measures for reducing risk and implement all the reasonably practicable ones, and dismiss the rest. Just remember to record what you did.

The ALARP Concept

So while the measures and 'sacrifice' required have the legal grounding in SFAIRP 'so far as is reasonably practicable', the risk side of the equation has come to be known by ALARP 'as low as reasonably practicable'. Note: that this isn't the UK legal standard, but it is used throughout the UK and wider world and has become an example of best practice.

The UK Health and Safety Executive gives an excellent description of what ALARP is actually used for and how it is implemented [HSE 2005];

> Using 'reasonably practicable' allows the legislators to set goals for duty-holders (the people in charge of the risk), rather than being prescriptive. This flexibility is a great advantage but it has its drawbacks, too. Deciding whether a risk is ALARP can be challenging because it requires duty-holders and the legislators to exercise judgement. In the great majority of cases, a decision can be made by referring to existing 'good practice' that has been established by a process of discussion with stakeholders to achieve a consensus about what is already considered ALARP. For high hazards, complex or novel situations, there is usually a need to build on good practice, using more formal decision making techniques, including cost-benefit analysis, to inform the judgement.

So for many of us, simply employing good practice will be enough to meet the ALARP (and hence SFAIRP) principle. However, there is a snag – how do you know what good practice is? Well, in the UK there is a definition, of course there is [HSE 2003];

> Good practice is the generic term for those standards for controlling risk, which have been judged and recognised by the UK HSE as satisfying the law when applied to a particular relevant case in an appropriate manner.
>
> Sources of written, recognised good practice include:
>
> 1. (i) HSC Approved Codes of Practice (ACoPs)
> 2. (ii) HSE Guidance.
>
> NB: ACoPs give advice on how to comply with the law; they represent good practice and have a special legal status. If duty-holders are prosecuted for a breach

of health and safety law and it is proved that they have not followed the relevant provisions of the ACoP, a court will find them at fault unless they can show that they have complied with the law in some other way. Following the advice in an ACoP, on the specific matters on which it gives advice, is enough to comply with the law.

Other written sources, which may be recognised, include:

Guidance produced by other government departments;
Standards produced by standards-making organisations (e.g. BS, CEN, CENELEC, ISO, IEC);
Guidance agreed by a body (e.g. trade federation, professional institution, sports governing body) representing an industrial/occupational sector.
Other, unwritten, sources of good practice may be recognised if they satisfy the necessary conditions, e.g. the well-defined and established standard practice adopted by an industrial/ occupational sector.

Demonstrating ALARP

Guidance on demonstrating ALARP is given in the UK by both military [MoD 2004] and civilian organisations [HSE 2003]. The MoD guidance on demonstrating ALARP suggests the following principles:

- Show that the sum of all risks from the system is in the broadly acceptable class.
- If that isn't possible, then show that the risks are not in the intolerable class.
- Identify risks that may be addressed by the application of good practice.
- Those risks not addressed by good practice need risk reduction techniques applied.
- Cost benefit analysis needs to be undertaken, with costs balanced against loss prevention.
- A grossly disproportionate factor of costs should be applied according to risk level.

The UK HSE guidance [HSE 2003] concentrates on the design phase, where it feels there is maximum potential for reducing risks. Their design guidance indicates the following principles:

- Carry out risk assessment and management in accordance with good design principles.
- Risks should be considered over the life of the facility and all effected groups considered.
- Use appropriate standards, codes, and good practices with any deviations justified.
- Identify practicable risk reduction measures and their implementation, unless the implementation is demonstrated as not *reasonably* practicable.

The HSE also makes some important statements of principle when considering ALARP [HSE 2001];

The zone between the unacceptable and broadly acceptable regions is called the tolerable region. Risks in that region are typical of the risks from activities that people are prepared to tolerate in order to secure benefits in the expectation that the nature and level of the risks are properly assessed and the results used properly to determine control measures; the residual risks are not unduly high and kept as low as reasonably practicable (the ALARP principle); and the risks are periodically reviewed to ensure that they still meet the ALARP criteria, for example, by ascertaining whether further or new controls need to be introduced to take into account changes over time, such as new knowledge about the risk or the availability of new techniques for reducing or eliminating risks.

The Accident Tetrahedron

Many texts cite accidents as occurring when people suffer an exposure to hazards – indeed this is the basis for some of the legal tests for negligence – potential harm (hazards), suffered loss (impact on the person) and causation (the exposure). It should be noted that the definition of accident might also include equipment, valuable assets, societal assets and the environment etc. This citation of exposure to hazards may be used as a generic approach to assessing accidents and their prevention. More importantly, the tool provides a route to demonstrate ALARP and SFAIRP arguments in a systematic way, and can be used to provide a route to proposing a justification for an argument about the completeness of the justification.

Developing parallels with the well understood combustion triangle (oxygen, fuel and heat) and fire tetrahedron (system reaction) [Sutton 2004], indicates that accident initiation should be prevented if;

> Potential exposure is sufficiently reduced in either the temporal or spatial frame of reference (so that vulnerable entities and hazards are prevented from sharing the same space and time, for example personal protective face equipment when operating a cutting machine).
> The number of people or value of assets used is sufficiently reduced (such that the consequences are below the threshold of a system's definition of accident, for example that the destruction of a remotely operated vehicle may be regarded differently to the destruction of a human operator).
> The severity of the hazard is sufficiently reduced (to a level that is tolerable and accepted, for example a lower strength or non-toxic material is used in some operation or system).

> and also the accident propagation should be prevented if;

> The reaction chain within a system can be broken (deliberate and planned intervention as an accident sequence develops to halt or reduce the progress of the event) [Maguire 2006].

This tetrahedral arrangement may be shown in a useful diagram, see below.

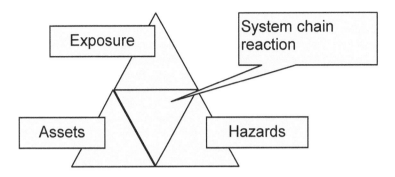

The original, specific fire tetrahedron indicates that if any of the elements is sufficiently reduced the propagation of the fire ceases and the full flame phase of the fire is prevented. One can envisage fire safety case reports using the fire tetrahedron as a basis for each operational mode of, for example, an off-shore fire management system, where the occurrence and significance of each tetrahedral element is judged to determine a level of risk priority. In turn each operational mode of a system is assessed to see if one or more of the four tetrahedral elements can be removed completely, or reduced to a level that would prevent fire propagation.

This is in fact what happens in many health and safety fire assessments – the necessity and quantity of fuel stored is reviewed and reduced, rich sources of oxygen are heavily controlled, and sources of heat, sparks or naked flames are prohibited. Chemical chain reactions are more complicated to control, but protective atmospheres and drenching systems can be used, although they may also impact on the quantity or dilution of oxygen present as well. As is often the case, one prevention method can often have an influence on more than one of the essential elements.

Using the developed accident tetrahedron and similarly focussing on preventing each component from being of sufficient size or magnitude to allow initiation to progress, can lead to a series of assessments that can be used in ALARP and SFAIRP justifications. For a complete justification, it will also be necessary to show that no further alternative mitigation or 'sacrifice' is justified on effort/cost/disruption grounds. This tool and method allows the user to develop convincing evidence that all strands of possible mitigation have been considered [Maguire 2006].

Problems with ALARP as a Safety Target

Whilst fairly well accepted in the UK and known of in Europe, USA and Australasia, ALARP is seen by many international viewers as difficult to verify and there is the perception that claims that all practicable risk reduction has been done, may be made without an appropriate effort [Bibb 2005]. Recent developments in The Medical Device Risk Management Standard (ISO 14971) indicate that the consideration of ALARP may be deleted from the new second edition due to the completeness question [Bibb 2005].

Discussions concerning the problem of demonstrating ALARP, have even reached the UK parliament. Medical discussions have highlighted that a risk appraisal process that excludes what are held by some stakeholders to be important factors, may fail to secure the crucial property of stakeholder confidence. It follows from this that, by instilling a misleading impression of completeness, robustness or rigour, risk assessments based on such incomplete risk characterisation may leave regulators and business highly exposed to a subsequent backlash on the part of the excluded parties [GeneWatch 1999].

In addition to the above there is a point of law currently (2006) being tested in the European Court of Justice. The problem, in short, is that European law imposes close to an absolute duty upon employers to safeguard their employees' health and safety. Whereas, in the UK, the HSE requires employers to safeguard health and safety 'so far as is reasonably practical' (SFAIRP). This is a classic example of 'goal-based' regulation type, of which the UK is justifiably proud.

The benefits of a goal-setting approach supported by case law are valuable. It gives flexibility to the UK law and permits a large degree of self-regulation. This contrasts with the continental codified system with its reliance on detailed comprehensive regulation. The two systems do not mesh very well and have created a number of difficult integration problems, most obviously seen in the 'so far as is reasonably practicable' issue.

The Commission appears to believe that the UK are trying to wriggle out of the responsibilities imposed by relevant European law, and have referred the UK to the European Court of Justice citing under-implementation of a 1989 EU directive on Health and Safety.

Real Use of the ALARP Process in Industry

In spite of the difficulties noted above, ALARP is used fairly extensively, particularly in the UK. One published safety case by London and Continental Stations & Property Limited [LCSP 2005] contains real use of the ALARP principle and process. This safety case applied the HSE guidelines on the tolerability of risk as a framework for assessing the acceptability of safety performance. These were used for defining the intolerable and broadly acceptable boundaries for the ALARP principle.

The ALARP assessment followed the stages described below [ibid.]:

1. The risk review group identified the various hazardous events associated with the operation of the station through a process of brainstorming
2. After this, the group went on to distinguish the precursors that could singly, or in combination with other precursors, cause the hazardous events to occur
3. The group considered all foreseeable failure modes that could lead to the occurrence of the hazardous events
4. The group considered any human factors related to the operation
5. The group considered local factors such as site/location specific features
6. The group considered the control measures already in existence

7. The group used their professional judgement to determine whether all reasonably practicable measures had been taken to reduce the frequency and consequence of each risk outcome
8. Through a process of brainstorming, the group identified any additional control measures required
9. Cognisance was given to the probable cost of implementing the additional measures and the fact that [other] work, currently underway, may well [already] encompass such measures.

This is a perfectly adequate risk assessment process for a safety case, and it contains a number of interesting points. The risk assessment team explicitly says that they used their 'professional judgement' in the analysis – excellent, very often this is actually missed out. The process also states that they considered the 'cost of implementing' the potential control measures that they had identified. Again very good.

Potential concerns are that whilst human factors get a special mention (good), mechanical, software and managerial factors do not. This implies some special attention to human factors. Also, it is not clarified what the 'cost' actually contains – it should contain factors relating to time, resource and trouble, not just a financial consideration.

One other detail that is missing (this may be due to commercial confidence and/or political aspects) is the value of the cost actually used in the ALARP decision, i.e. what is the cost level where the company will have to apply additional mitigation. This is a complex and ethical question, which quickly boils down to the concept of a value for preventing an accident or a fatality. It is not surprising that it is not freely quoted. This concept will be considered later in this book.

The GALE Principle

In most instances in the UK, ALARP or SFAIRP will be the over-riding concepts for risk assessment and safety case use. This is because of the legal position of the terms, and their uses in UK Health and Safety at Work legislation. In other countries and in non-work domains, a different risk principle may be applied (unless specific legislation and/or best practice dictate otherwise). One alternative principle is GALE.

In this instance GALE stands for 'globally at least equivalent'. Under this regime, if there are circumstances (change, modification, employee reduction, whatever) where the risk associated with a particular hazard increases, the whole system will remain risk acceptable if it can be shown that risks from other hazards have been reduced by an equivalent or greater amount. The principle does not remove the need to carry out risk assessments and record them, nor does it discharge you from applying risk mitigation measures where reasonable to do so. It does exempt you from proving that the risks from your system is *as low* as reasonably practicable – they just have to be lower than before the change.

This principle has been applied to a road traffic improvement programme in the UK [Halbert & Tucker, 2006]. It was justified as appropriate in that the

Health and Safety at Work Act did not place a duty on the UK Highways Agency to achieve ALARP as far as members of the public using the road were concerned, only on the employees constructing the road improvement.

The implications of using this principle as a global safety target are difficult to evaluate and may potentially give rise to some serious questions, if considered on a national basis. One example where this type of risk evaluation may be critical is when you consider the whole transport infrastructure of a country [Goosens & van Gelder, 2002]. For example, suppose major repair activities on the railroad tracks require no train to run for a significant period of time. One of the main reasons for doing so is given by the relatively high individual risks of the rail workers during their construction tasks. Shutting down all train activities along the repair tracks reduces this risk to zero far better than any other decision with trains still being allowed, but against certain constraints like lower speeds and lower frequency.

However, the no train at all decision requires alternative transport modes, like transporting train passengers with buses. Furthermore, a large part of the passenger population will make a decision to go by car. So the safety decision in the rail system creates higher risks in the road system. As the rail system is generally safer than the road system, we may eventually end up with an overall higher risk profile for national transport. A difficult decision has to be made, in trading off the rail worker risk against increased road traffic risk to maintain an at least equivalent risk profile.

Notes

Bibb 2005: "The Medical Device Risk Management Standard – An Update.", The Safety-Critical Systems Club Newsletter, Vol. 14 No. 3, May 2005.

Court of Appeal 1949: "Judgement in Edwards v. National Coal Board", [1949] 1 All ER 743.

GeneWatch 1999: "Appendix 32 to the Minutes of Evidence II, Scientific Advisory System : Genetically Modified Foods". Select Committee on Science and Technology, House of Commons, May 1999.

Goosens & van Gelder 2002: "Fundamentals of the Framework for Risk Criteria of Critical Infrastructures in The Netherlands". Elsevier Science Ltd. 2002.

Halbert & Tucker 2006: "Risk Assessment for M42 Active Traffic Management", Safety Critical Systems Club, Springer, 2006.

HSE 2001: "Reducing Risks, Protecting People – HSE's Decision Making Process." HMSO Norwich, 2001. ISBN 0-7176-2151-0.

HSE 2003: "Assessing Compliance with the Law in Individual Cases and the Use of Good Practice", Health and Safety Executive, 2003.

HSE 2005: "ALARP at a glance", Health and Safety Executive, 2005.

LCSP 2005: "St. Pancras Station Railway Safety Case", London & Continental Stations & Property Limited, London, May 2005.

Maguire 2006: "So how do you make a full ALARP justification? Introducing the Accident Tetrahedron as a guide for Approaching Completeness", Safety Critical Systems Club, Springer, 2006.

MoD 2004: "Safety Management Requirements for Defence Systems Part 1" Interim Defence Standard 00:56, Issue 3. Ministry of Defence, December 2004.

Sutton 2004: "Notes for Guidance", The Fire Safety Advice Centre, Merseyside Fire Liaison Panel, 2004.

Chapter Ten

Individual, Group and Population Risk

Sharing Risk

Risk may be distributed and shared amongst, or shared between a number of persons. This chapter discusses understanding the way risks may be different if you are an individual person, a small group or a whole population. For a quick example, consider an explosives disposal unit carrying out their 20 or so defuse operations per year. In the UK, these teams operate around the coastline dealing with washed-up unexploded or dangerous-looking devices. The single lead person who initially has to approach the explosives is taking the whole risk for the group – but if the team take it in turns to defuse each bomb, then they are sharing the personal risk. However, the whole group risk remains the same.

Historical Incident

A man has been discharged from hospital after being seriously injured while testing a rifle-fired grenade at a military range in West Wales. It is understood the employee was injured in his arm and abdomen by flying shrapnel at the Pendine Ranges in Carmarthenshire. The incident, which happened at 1345 BST on Tuesday 15th April 2003, is being investigated by the Ministry of Defence (MoD), the Health and Safety Executive and Dyfed-Powys Police[BBC 2003].

Individual Risk

As we have seen, rates of incidents (targets or actual data) are often expressed as the likelihood of a particular outcome per 'x' number of events, even if those events are just days of operation i.e. in the time base. When a rate is expressed as the likelihood of a fatality per 100,000 worker days of labour, this has a number of meanings. For example, if you had 10 people working for 10,000 days (a 45 year working life) in a particular occupation where the historical fatality rate was one fatality per 100,000 worker days, this rate would be correct if one of the ten was killed during his working life, his death being caused by something at work (not an event away from work).

However, if you had a company of 1,000 staff, it might be said that there would be one fatality every 100 days. This appears to be very shocking and far too high. In both cases the individual rate is still 1 per 100,000 worker days, but the group risk is very different – 1 per 45 years or 1 per 100 days. The data only makes real sense where the risks to the individuals are gained from the historical data of risks to the whole industry population.

Consider the following data from the US Bureau of Labor Statistics [US DoL 2005] and the UK Health and Safety Commission [HSC 2005a], which quote actual recorded data for workplace fatalities in a number of industry types;

Table 10.1 US Industry Fatality Statistics

US Industry	Number of fatalities (Y2004)	Workplace fatality rate per 100,000 workers (Y2004)
Manufacturing	459	2.8
Construction	1224	11.9
Mining	152	28.3
Government	526	2.5
Agriculture	659	30.1
Transport & Warehouse	829	17.8
Retail Trade	372	2.3
Leisure	245	2.1
Finance	115	1.2
	ALL INDUSTRY	4.1

Table 10.2 UK Industry fatality statistics

UK Industry	Number of fatalities (FY2004/05)	Workplace fatality rate per 100,000 workers (FY2004/05)
Manufacturing	41	1.2
Construction	72	3.5
Extraction and utility	2	1.1
Service industries	63	0.3
Agriculture	42	10.4
	ALL INDUSTRY	0.6

NB. As this data is for a 12-month period, the values also represent the fatality rate per worker-year, as the denominator of the equation would be 1. It should also be noted that each country does not necessarily define the industry sectors in the same way, however, the definitions are considered close enough to be useful for general comparison – particularly the all-industry figure.

Group Risk

Consider any one of these industry sectors, let's pick on US construction. From the data we can see that there were 1224 fatalities in 2004 at a rate of 11.9 per 100,000 workers over the working year. This is roughly 0.1% or 1×10^{-3}, so you can say that each individual in the US construction sector has a risk of fatality of

0.1% per year. For a large construction company employing a group of workers that numbers 1000 or so, the probability of a fatal accident during the year is getting close to 1, i.e. a certainty. One of the staff at work will be killed by an accident at work over the next year. So the risk exposure for the individual might be considered as fairly low, however, the risk exposure for all the workers exposed to the same risk is much higher.

Now let me ask you, what do you think about these numbers? Is a group risk of fatality at 1 really tolerable? Discuss.

After some thought, you might consider that these fatality rates are pretty much tolerable, after all nobody has stopped all construction work in these countries – the economic consequences would have been catastrophic (therein lies a classic example of risk vs. benefit in real life). However, legislators in these countries have indeed decided that these fatality rates are not tolerable – targets for reduction have been developed and published.

In the UK a document was published in 2000 on the subject of revitalising health and safety in the workplace [Department for Environment, Transport and Regions 2000]. This gave targets for fatality reduction, which might be viewed as something approaching national safety targets for use in a hypothetical national safety case (!):

- Reduce the number of working days lost per 100,000 workers from work-related injury and ill-health by 30% by 2010

- Reduce the incidence rate of fatal and major injury accidents by 10% by 2010

- Reduce the incidence and rate of cases of work-related ill-health by 20% by 2010

- Achieve half the improvement by 2004.

The published statistics on UK health and safety at work inform the measurement of progress against the targets for reducing work-related injuries, ill health and working days lost set in the 'Revitalising Health and Safety' strategy. Annual progress reports have been published each Autumn since then. (All these documents are on the HSE website at www.hse.gov.uk/statistics/targets.htm.) HSE statisticians' latest assessments, at the mid-point of the strategy period, are as follows [HSC 2005b]:

Progress on fatal and major injuries

The Revitalising Health and Safety target for 2004/05 is to reduce the incidence rate of fatal and major injury by 5% from 1999/2000. The available sources indicate no clear change since the base year in the rate of fatal and major injury to employees. The target has therefore not been met.

Progress on work-related ill health incidence

The Revitalising Health and Safety target for 2004/05 is to reduce the incidence rate of work-related ill health by 10% from 1999/2000. The evidence suggests that incidence has fallen for most major categories of work-related ill health. Overall, the 10% target has probably been achieved.

Progress on working days lost

The Revitalising Health and Safety target for 2004/05 is to reduce the number of working days lost per worker due to work-related injury and ill health by 15% from 2000-02. There has been a significant fall in working days lost since the base period, possibly enough to meet the 15% target.

So whilst some great progress has been made, there is still critical work to be done regarding the incidence rate of fatal and major injury. It is difficult to see how this is to be done, the first five years has not shown any clear change since the base year – what is to be done differently over the next five years? A significant available strategy is to make having a fatal or major accident very financially harmful, perhaps by increasing fines and other legal and operational penalties. Hence industry will not want to have this type of event, and hence more effort will be focussed on systematically assessing safety, risk and hazards – more safety cases.

In the US, the pattern is pretty much the same, the OSHA strategic management plan [OSHA 2003] specifically cites;

> Since OSHA was created in 1971, the workplace fatality rate among employees has decreased by 62% and occupational injury and illness rates have declined by 42%. At the same time, US employment in the private sector as well as the number of workplaces has doubled, increasing from 56 million workers at 3.5 million establishments to 114 million workers at 7 million establishments. The decrease in fatalities, injuries and illnesses across such an expanding population of workers demonstrates remarkable progress. Nevertheless, the number of reported fatalities, injuries, and illnesses *remain unacceptably high*. In 2001 there were 5,270 fatalities in private industry and in 2000 there were more than 5.7 million reported injury and illness cases.

In response to these figures being unacceptably high, perhaps intolerable, a number of performance goals have been set to be achieved by 2008. These are to be tracked according to the US Government Performance and Results Act requirements. The two primary goals are [OSHA 2003]:

3.1C   By 2008, reduce the rate of workplace fatalities by 15%
3.1D   By 2008, reduce the rate of workplace injuries and illnesses by 20%.

Whilst these are not legislative in nature, the UK and US goals will be seen as industry and government supported guidance and requirements. Any quantitative safety targets that are used as part of a safety case should bear these in mind, and reference them as influencing factors.

Population Risk

To evaluate risk for larger groups or a whole population, the whole risk and the nature of the distribution needs to be understood. Is the risk shared equally, or are there factors affecting the loading of risk across the members? A method for assessing the risk over a population or group needs to be established and agreed.

One way of regarding population risk is to understand the relationship between the frequency of occurrence and the number of individuals suffering from a specified level of harm within a given population, from the realisation of a specified hazard. An accepted method of expressing this relationship is through a 'Frequency vs. Number of fatality' graph, known as 'FN' curves. This graph can be used to represent lines of tolerable and broadly acceptable risk levels for increasing proportions of the population. A logarithmic scale is usual (but note, not necessarily to the base of 10), with each factor of increase in the number of the population exposed to a risk, there is a required decrease in the occurrence of that risk. A typical FN curve is shown on the next page. The gap between the two lines is the area where the concepts of ALARP and SFAIRP would come into play – a tolerable risk, if the benefit were worth it.

The FN curve has been used extensively in The Netherlands to assess the potential and tolerability for flooding in the large areas of the country that are below sea level [VROM 1988]. Since its use, a great many additional factors have come to light about risk interpretation that may be gained from them. For example, the gradient of the curves themselves are seen as a measure of risk aversion or acceptance – the steeper the gradient the more risk averse society is concerning that particular hazard. Figure 10.1 has the boundary lines arranged in a linear fashion, the actual intolerability line for society is more likely to be exaggerated at the higher fatality numbers. The 'real' societal opinion lines are likely to be more like an arc, horizontal at the top and with a vertical cut off to the right. This is shown in Figure 10.2.

The area under the lines to the x-axis can be used to express a measure of the tolerated and not-tolerated societal risk. It can also be shown mathematically that the area corresponds to the expected value of the number of deaths [Vrijling & van Gelder, 2000].

Australia has also used FN curves for developing societal risk criteria. New South Wales planning department and the Victoria Work Cover Authority have published specific FN curves for their states [Robinson et al, 2006] . Here, it has been noted that once a society-determined death threshold has been passed, it appears that the community has a much greater aversion to single incidents with increasing fatalities. This gives some credibility to the idea that the lines on the FN graph need not necessarily be straight.

A Note about the Examples

Please remember, these examples are just that – examples, you should not copy them directly for use in your particular project. Your own FN curves may end up looking similar, but you will need to derive them yourselves from your own judgement or researched evidence of how society views risks from your project.

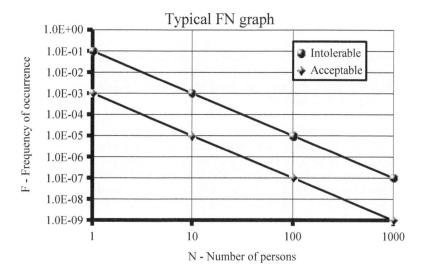

Figure 10.1 Typical FN Graph

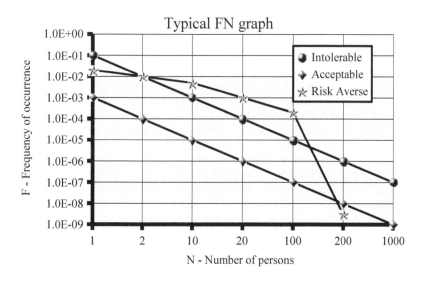

Figure 10.2 Typical FN Graph with Possible Real Risk Aversion Factor

Use of FN Curves

FN curves are specific to a particular event outcome, e.g. loss of life, loss of environment, injuries, even loss of data or equipment. So several may have to be drawn up for a project for each domain of risk you are considering in your safety case. Don't just think that they are confined to human population deaths, they can have wider use if thought about.

A simple method for deriving a usable FN curve is to identify the level of probability of a single fatality that is tolerated in society from a similar project or system to yours. This can be done by reference to existing tolerability data for a similar project or system, or by reference to guidance from a national regulator, or from your own empirical research. The next step is to repeat the exercise but for a fatality level of 1000 persons. The final step is to represent these on a graph and join them up with a straight line. You might decide that the curve should be biased at the high end – that is your professional judgement, no one else can do that for you.

This method should be repeated for levels of probability that are considered intolerable in society, leading to a pair of FN curves for your project or system. It is not necessary that the two lines are parallel, it just means that the ALARP area is perceived differently at different levels of risk – after all there is no reason why it should be treated the same.

However, it should be noted that the effort to ascertain very accurate representations of FN curves needs to be balanced against the benefit of use that is to be gained from knowing them in that much detail. It may not be worth developing the graph any further than as a couple of straight lines.

Worker vs. Public Risk

The workers and public that are the main groups assessed for risk exposure are usually treated as first and third parties with respect to the system under analysis. There is a distinct difference in the handling of risk exposure to these two groups.

First parties are involved directly with the system and they get some benefit from doing so – job satisfaction, pay and (sometimes) a pension. They have implicitly accepted the risk of doing their jobs in order to receive some benefits. In fact they have accepted the whole risk profile of their employment role, not just the obvious heavy, sharp, toxic parts, but also the noise, stress and electromagnetic parts too. Indeed part of the role of the safety case is to assess all these areas and demonstrate that the risks to first, second and third party humans (and the environment etc) from these sorts of effects are really at a tolerable level and ALARP.

Third parties have no involvement at all with the system under analysis – they probably do not even know of its existence. They get no direct benefit from it, so why should they have to accept any risk from its use? This is a strong argument to break. It may be argued that the third party general public gets some implied benefit from a particular system or piece of new equipment – it may contribute to national security or national economy or some other wider societal interest.

However, once this has been diluted down over the number of people in a given population, the individual benefit tends to the negligible. Third parties do not want to accept any risk from something they've never heard of – the public can be very risk averse.

There should now be a question on the tip of your mind – if these groups have different risk tolerability levels, just how different are they? How much (if any) extra protection should be applied to the public over the worker?

In the UK, the HSE has produced guidance [HSE 2001] on reducing risks and protecting people. Within this document, the HSE discusses the boundary between unacceptable and intolerable, it states;

> ... we suggested that an individual risk of death of one in a thousand per annum should on its own represent the dividing line between what could be just tolerable for any substantial category of workers for any large part of a working life, and what is considered unacceptable for any but exceptional groups. For members of the public who have a risk imposed on them 'in the wider interest of society', this limit is judged to be an order of magnitude lower (i.e. $1/10^{th}$) – at 1 in 10,000 per annum.

As far as Europe as a whole is concerned, there is no pan-Europe legislation existing to give guidance on what level of third party risk is acceptable [Eurocontrol, 2006]. However, the concept of Public Safety Zones around airports (and other hazardous installations) has been developed. This is where the probability of fatality in society around a hazardous site is calculated in probability zones – higher, nearer the risk source and lower, the further away. It has been recommended that new public housing should not be permitted within the PSZ contour for a fatality risk of 1×10^{-5} per annum, and that dwellings (already) within the 1×10^{-4} contour should be purchased by the site operators and the occupants moved away.

In the US, the concept of societal risk reduction has a political aspect, related more to a social health responsibility rather than risk exposure from equipment and industrial systems. As in Europe, there appears to be no legislation giving guidance on third party risks. At this particular time, concentration is on homeland security and protection from external threats, rather than internal hazards.

Multiple Safety Targets in a Safety Case

The safety case should develop a scale for assessing the safety performance of the equipment or system of interest. The scale should enable all stakeholders to understand the level of risk that they may be exposed to. This means that a safety scale has to be developed for each stakeholder group.

It may be that a single scale might be applicable to all stakeholders, but this is considered unlikely, as there is certainly the case for the general public to be exposed to a lesser risk. The implication of this is that multiple safety scales – and therefore multiple safety targets – have to be defined. The equipment or system should then be assessed against all of them. The following spread of safety scales is highlighted:

- For individual, first party workers
- For the group representing second party co-workers or site visitors
- For individual third party members of society
- For third party population groups.

These may also have to be multiplied for different unit types, if you are considering a situation similar to the transport case with three different unit types in common use.

So instead of just having a single safety target to be assessed against, there should actually be several. They may be further safety scales related to environmental impact, financial impact and any other domains of interest. You will be lucky if there is just a single scale and target to consider in the safety case.

Notes

BBC [2003], "Man injured in grenade blast", BBC News, Cardiff, 2003. http://newswww.bbc.net.uk/1/hi/wales/south_west/2951999.stm

Department of the Environment, Transport and the Regions [2000], "Revitalising Health and Safety – Strategy Statement June 2000", Department of the Environment, Transport and the Regions, London, 2000.

Eurocontrol [2006], "Environmental Issues for Aviation – Third Party Risk", Eurocontrol, Brussels, Belgium, 2006.

HSC [2005a], "Statistics of Fatal Injuries 2004/05", The Health and Safety Commission, Bootle, UK, 2005.

HSC [2005b], "Health and Safety Statistics 2004/05", The Health and Safety Commission, Bootle, UK, 2005.

OSHA [2003], "Occupational Safety and Health Administration (OSHA) 2003-2008 Strategic Management Plan", US Department of Labor, 2003.

Robinson et al [2006], "Risk and Reliability – An Introductory Text" 6th Edition, R2A, Melbourne, Australia, 2006.

US DoL [2005], "Census of Fatal Occupational Injuries (CFOI) – Current and Revised Data", US Department of Labor, Bureau of Labor Statistics, 2005.

Vrijling & van Gelder [2000], "Societal Risk and the Concept of Risk Aversion", Delft University of Technology, Delft, 2000.

VROM [1988], "Dutch National Environment Plan", Netherlands Ministry of Housing, Spatial Planning and the Environment, The Hague, 1988.

Chapter Eleven

The Safety Team

Why Have a Team at All?

Over the first ten chapters of the book a great deal of analysis and assessment work has been written about – hazard analysis, process analysis, matrix development, target setting, interpreting regulations etc. Just a few people with plenty of time can do all this. But that's the point, whilst there may be just a few people who can/want do this, there probably isn't the time. Further more, there probably isn't all the appropriate knowledge and experience condensed into just a few people. There will need to be a group of people involved, and for any group effort, it is worth having the group working as a team.

As a demonstration of the breadth of experience needed, below is the list of the group brought together to carry out the risk assessment on the St. Pancras Station Railway Safety Case in the UK [LCSP 2005];

- Station & Safety Manager
- Compliance Manager
- Station Project Manager
- Duty Station Manager
- Risk Analyst
- Human Factors Specialist
- Head of Safety
- Interface Manager
- Contractor's Design Manager.

There are two important things to notice about this group: one, look at the seniority of the people involved – Heads and Managers. This is good. Two, with regard to the seniority of this group, think about how expensive a single meeting would be between them. This is not so good. As the counter to the second point though, think also about the expense of missing one day's operation through having a minor accident, or the expense of having a multiple fatality train crash in the station. Easy decision to have the meetings really.

This example is just the risk assessment group, responsible for carrying out the hazard, risk, and therefore, safety analysis. There is a much wider team involved in the whole safety management system, including (but not limited to...) safety managers, safety engineers, domain specialists, system engineers, customers and regulators.

What the Team has to Do

Ultimately the team has to put together the safety case and write the safety case report. All the target interpretation, risk analysis, testing, explanation and evidence-collecting has to be done by this team. But whole team responsibilities go much further than this.

Typically someone will have to be responsible for (in no particular order):

Appointing the safety engineering staff and team members
Authorising and approving the Safety Management System
Reviewing staff against competence criteria and managing training
Endorsing and accepting all safety cases (usually via a signature on the cover)
Managing the system safety group infrastructure
Establishing and maintaining the system safety policy and procedures
Ensuring that satisfactory audits are carried out
Managing changes to the project and/or the status of the standards etc.
Establishing the design rules and techniques to meet the safety targets
Providing the funding for all safety activities identified
Ensuring the provision of sufficient schedule time and resources
Ensuring that all components are subjected to appropriate analyses
Ensuring that any hazards identified are entered in the records log
Executing the safety verification and validation activities
Implementing the safety tasks identified
Compiling the safety case report and safety evidence
Maintaining safety plans and other safety documents
Chairing the project safety meetings
Reviewing and approving any proposed design changes for their effect on safety
Reviewing the test programme to ensure that it adequately covers safety features
Reviewing the incidents arising during test and evaluation
Review and audit of the safety procedures
Review and audit of the safety documents
Providing regular status report to the Safety Manager and Programme Manager
Maintaining the system hazard log
Executing and recording system hazard analysis
Evaluating the product/system against all relevant national/ international legislation
Acting as archivist/secretary at all the safety meetings.

Referring back to the first section in this chapter, if someone out there thinks they can really do all this on their own, they should not be reading this book – they don't have the time!

Who is in the Team?

In a project of even a reasonable size a team or group of people will be required to carry out all the required safety tasks. A typical core safety team might be as follows;

Project Manager

This person should be the lead example for safety management informing the project of the importance of safety and detailing specific safety responsibilities to other project staff. The project manager should be the first point of contact for safety related issues. The project manager should manage and oversee the safety tasks, they should review all the safety reports, including the safety case and hazard log. This person should be given appropriate authority to enforce safety practices and to reprimand those who act in an unsafe way. The project manager should have greater than five years safety experience and be highly qualified in a related engineering subject – preferably of chartered status.

Project Safety Engineer

This engineer should be responsible for the day-to-day implementation of the safety tasks that are required by the project. On larger scale projects, this person may be in charge of a number of people across several areas of specialism in the project. The project safety engineer is usually the prime author of the safety case report, but naturally contributions will come from other sources. The project safety engineer should have greater than two years safety experience and be highly qualified in a related subject – preferably of chartered status.

Project Safety Committee Members

The committee should be viewed as a forum for implementing the safety requirements into the functional design of the system. The project safety committee should be established at the start of the project. Members should be made up from internal company departments – design, manufacturing, senior management, regulators/auditors and some external bodies – customers and contractors.

Subject Matter Experts

On many projects, there will be areas where detailed knowledge is required about the operation, performance or characteristics of a particular part of a system. It may be that a particular safety assessment method is required, where the company does not have enough detailed knowledge. In these circumstances, one or more subject matter experts may be drafted onto the safety committee for specialist advice.

Independent Safety Auditor

The overall objective of the independent safety auditor is to ensure that the safety issues have been addressed via proper application of the appropriate standards and legal requirements. The exact scope of the auditor's work should be confirmed in an audit plan that is agreed to by the project manager and project safety engineer. It is likely that this person will audit the safety case report and other safety related documents, and also review project procedures for managing safety in the future.

The Project Safety Committee

At regular intervals during a project, the safety team should get together to discuss the safety and risk performance of the project as a whole. This meeting is generally known as the project safety committee, or it might be the project safety working group or the safety team committee. In larger projects, there may actually be a hierarchy of groups, with each having a different range of responsibilities and authorities.

There are several ideas about what a safety committee is in place to do, for example from the UK defence industry [MoD 2004]:

> The safety committee shall oversee, review and endorse safety management and safety engineering activities ... the safety committee shall make recommendations ... on safety matters that require formal agreement before proceeding e.g. ALARP judgements.

But around the world there are multiple and various examples of this sort of committee, carrying out specific roles in specific industries, and at specific levels of responsibility:

> A joint health and safety committee (JHSC) is a forum for bringing the internal responsibility system into practice. The committee consists of labour and management representatives who meet on a regular basis to deal with health and safety issues. The advantage of a joint committee is that the in-depth practical knowledge of specific tasks (labour) is brought together with the larger overview of company policies, and procedures (management). Another significant benefit is the enhancement of a co-operative attitude among all parts of the work force toward solving health and safety problems. In smaller companies with fewer than a specified number of employees a health and safety representative is generally required. Consult health and safety legislation for details.
>
> The committee may also be known as the Industrial Health and Safety Committee, Joint Work Site Health and Safety Committee, Occupational Health Committee, Workplace Safety and Health Committee, or Health and Safety Committee [CCOHS 2004].
>
> The National Marine Safety Committee (NMSC) is an Intergovernmental Committee established in 1997 ... to achieve uniform marine safety legislation and practices throughout Australia. The committee is comprised of senior executives of marine safety agencies throughout Australia, with the New Zealand Maritime

Safety Agency having observer status. The committee is supported by a small secretariat based in Sydney. The NMSC consults widely with industry and invites industry members to sit on a range of reference, advisory and professional panels. The Committee meets formally three to four times a year to review progress, set priorities, and to endorse the outcomes of the projects, and to provide recommendations to the Australian Transport Council (ATC) through the Australian Maritime Group [NMSC 2005].

An Independent Safety Committee shall be established consisting of three members, one each appointed by the Governor of the State of California, the Attorney General and the Chairperson of the California Energy Commission, respectively, serving staggered three-year terms. The Committee shall review Diablo Canyon nuclear power plant operations for the purpose of assessing the safety of operations and suggesting any recommendations for safe operations. Neither the Committee nor its members shall have any responsibility or authority for plant operations, and they shall have no authority to direct Pacific Gas and Electric Company personnel. The Committee shall conform in all respects to applicable federal laws, regulations and Nuclear Regulatory Commission ('NRC') policies [DCISC 1988].

Forming a Safety Committee

Excellent guidance is given on the formation of a safety committee by OSHA in Oregon in the US [OR-OSHA 2006]. They propose that the purpose of a safety committee is to bring workers and managers together to achieve and maintain a safe and healthful workplace [or project]. To set up a safety committee, they have provided a very useful taxonomy [ibid].

- Decide if your company needs a safety committee
- Review OSHA's safety committee rules [terms of reference (TORs)].
- Understand a safety committee's seven essential activities
- Determine where you need a safety committee
- Determine how many representatives will serve on the committee
- Determine who will serve on the safety committee
- Set practical goals for the safety committee
- Train safety committee representatives
- Hold regular safety committee meetings.

You will have noticed the hook for the seven essential activities, and I know you are desperate to know what they are [ibid.].

- Gain management commitment
- Be accountable for achieving goals
- Involve employees in achieving goals
- Identify workplace hazards and make recommendations on their control
- Review reports of accidents and near misses
- Keep accurate records of committee activities
- Evaluate safety strengths and weaknesses.

These guidelines are generic enough to be used in most situations, and equally, they are detailed enough to be useful in most situations.

Who Owns the Safety Case?

An essential concept for a safety case is to understand who owns it – well the real question is not who owns the *safety*, but who owns the *risk*? The main principle being that whoever owns the risk should own the safety case. Unfortunately this is not always the case.

It is critical to the success of a safety management system (and hence the value of a safety case) that the ownership of safety is clearly understood. If there is no owner, the concept becomes derelict and ruined. In the UK the Health and Safety at Work laws mandate a duty of care on employers, employees and everyone else who comes in to contact with the hazardous system. There is also the duty on everyone else including visitors, contractors and even trespassers, not to interfere or mis-use any safety equipment provided in order to enhance safety in the workplace.

This does mean that each of those groups can be prosecuted for transgressing these laws. This includes company executives for allowing unsafe procedures to be in existence (even if no-one has actually been harmed), and trespassing juveniles who have broken into a factory site and set off the fire extinguishers as a prank.

So there is the argument that everyone has a responsibility for safety, and therefore everyone owns the safety case – but of course not all these groups have authority to manage and affect safety. It is the seat of authority that also holds the ownership of the safety case. In most companies, the directors and executives have the authority of managing their company, their contracts of employment will give them power to run and influence the company.

It is likely that there will be a nominated 'safety manager', either for the whole company or, in very large organisations, several safety managers spread across multiple projects. This person may well have official delegated authority for safety and may actually be known as the '*duty-holder*'. It should be this person's responsibility to oversee the construction of a safety case and the evidence required to prove the safety claim. Of course, this person should also be given full authority to ensure that safety tasks and safety analysis are done. The duty holder has the role of ensuring that the people, products and processes fulfil their statutory and regulatory duty of care for safety.

Notes

CCOHS 2004, "What is a Joint Health and Safety Committee?", Canadian Centre for Occupational Health and Safety, Hamilton, Canada, 2004.

DCISC 1988, "The History of the Diablo Canyon Independent Safety Committee", DCISC, Monterey, USA, 1998.

LCSP [2005], "St. Pancras Station Railway Safety Case", London & Continental Stations & Property Limited, London, May 2005.

MoD 2004: "Safety Management Requirements for Defence Systems Part 1" Interim Defence Standard 00:56, Issue 3. Ministry of Defence, December 2004

NMSC 1997, "What is the National Marine Safety Committee?", NMSC, New South Wales, Australia, 1997.

OR-OSHA 2006, "Safety Committees for the Real World", (3/06/COM), Oregon Occupational Safety and Health Administration, Salem, USA, 03/2006.

Chapter Twelve

Costs in Safety

The Measurements of Costs

In the safety domain sooner or later you will come across expressions of costs, the costs of accidents, the costs of preventing accidents, the costs of producing a safety case etc. These are all different costs, even though they look like they might be the same. Or nearly the same at least.

The cost of an accident should comprise only the direct costs of the event itself, the material, equipment and vehicles lost, the cost of emergency service attendance and any environmental cleaning required. A further factor that might be considered here is the cost of any penalties or fines.

The cost of preventing an accident is different from the direct cost of an accident. It does not represent the actual costs incurred, it is the cost-benefit value and actually represents the benefits which would be gained by the prevention of the accident. This cost is better described as a value, and should incorporate such things as lost domestic product output, human or social factors in addition to the direct cost groups noted above.

The cost of producing a safety case is not related to having or not having an accident. You might think that having a safety case should prevent accidents, well in a sense this is true. There is evidence *of* safety cases reducing hazards and accidents, but it cannot be said that having a safety case is evidence *for* avoiding accidents. The development of a cost for producing a safety case will be dealt with later in this book, so for now the text will concentrate on the cost of having accidents and the value of preventing an accident.

The Cost of Having Accidents

The Health and Safety Executive has performed an excellent series of case studies on UK industry. Multiple industry sectors have been closely analysed over relatively short time periods and the accidents and incidents that occurred have been fully costed out to demonstrate the financial impact of having an accident.

Incidents at a National Health Hospital [HSE 2002a]

The study took place at a 367 bed hospital employing around 700 staff. The accident criterion for this study was any unplanned event that resulted in injury or ill-health to people, damage or loss to property, materials, products or the environment, or loss of opportunity to provide a service or treatment. The study lasted for 13 weeks. There were 1232 reported accidents, six over-three-day injuries, 19 other injuries to staff, 38 injuries to patients and one to a visitor.

Financial costs were   £63,224
Opportunity costs were   £66,045

Many of the losses were directly attributable to poor preventative maintenance of equipment, plant and buildings. 78% of the personal injury accidents occurred in the care for the elderly wards. On an annual basis, the total costs represented 5% of the hospital annual running costs.

Incidents on a Large Construction Site [HSE 2002b]

A supermarket was being constructed by a main contractor and 29 sub-contractors. The study covered 18 weeks spanning the ground works and roofing stages. The accident criterion used was any unplanned event that resulted in injury to people, damage or loss to property, products, material or the environment or a loss of business opportunity.

There were 3825 accidents including 56 minor injuries, there were no major injuries, dangerous occurrences (near-misses) or three-day injury accidents.

Financial costs were   £119,519
Opportunity costs were   £222,294

£13,175 of the financial costs were the result of lost materials, the total uninsured losses for the main contractor were £102,825. The contribution to the total costs for the actual 65 injury accidents was only £580.50.

Incidents at a Financial Administration Company [HSE 2002c]

The study was conducted at a financial services organisation involved in cheque clearance processes. This was an office environment where the main activities were working with display screen equipment, manual handling and clerical duties. The accident criterion used for this study was personal injury and loss. The costs also include the impact of providing cover for absent staff.

The study was over 13 weeks, there were nine injury accidents and four work-related ill-health absences.

Total cost was   £22,822

Seventy percent of the total cost was due to the ill-health events, which represented an average cost of over £3000 for each case. The total cost represented 0.5% of the annual salary bill.

Incidents at an Oil Production Platform [HSE 2002d]

This North Sea oil production platform was situated 100 miles from land and had up to 120 people staffing the rig, which operated on a continuous 2 x 12-hour shift pattern. The average production rate during the study was below 40,000 barrels per day, but the company considered that this represented a typical three month operating schedule. The accident criteria used was any unplanned event that resulted in injury to people, damage or loss to property, products, material or the environment, or a loss of business opportunity (non-production).

Over the 13 weeks, 262 accident events were recorded for the study, including two three-day personal injuries and eight other injuries requiring first aid.

Financial costs were   £1,141,586
Opportunity costs were   £   143,815

It was calculated that the total cost (post tax) to the company was equivalent to £1.80 per barrel of oil. Losses were equivalent to one day's lost production per week of operation.

Incidents at a Transport Company [HSEe 2002]

This study took place at a transport depot operating a 140-strong fleet of milk tankers and delivery vehicles. There was also a vehicle maintenance department on the site. The accident criteria used for this study was an unplanned event that resulted in injury to people, damage or loss to property, materials or products, or loss of business opportunity.

Over the 13 week study period there were no personal injury accidents, but a total of 296 other events within the definition occurred.

Financial costs were   £21,972
Opportunity costs were   £44,326

Commentary on Case Studies

The one thing that struck me as I was researching this topic was the sheer numbers of accident events occurring throughout industry – 200 per week on a single large construction project. Admittedly, these are all events coming within the definition of accident used for the study, but it seems incredible how many there are and how expensive they are to each industry and therefore to a country's economy as a whole. Interestingly, the costs identified in just these studies works out at around £320 per accident. That doesn't sound much, but when you might be having between 5000 and 10,000 per year, that is *very* expensive.

It is anticipated that these values show an indicative broad status of the cost of having accidents across a variety of industries. It is also anticipated that these values would be fairly consistent across the industrialised world.

The Value of a Prevented Fatality

In the US, the National Safety Council makes an estimate of the average costs of fatal and non-fatal injuries from unintentional sources as a way to illustrate the impact of these events on the US national economy. The costs are a measure of the money spent and the income not received due to accidents, injuries and fatalities. It is regarded as a further way to understand the importance of preventative measures. The determined values can be used to estimate the financial impact on a state or local community, and can be used for cost-benefit analysis. [NSC 2005].

Costs of Motor Vehicle Injuries

> This cost is calculated from an assessment of wage and productivity losses including the total of wages and benefits; medical expenses including doctor fees, hospital fees, costs of medicines and emergency medical vehicle services; administrative costs including public and private insurance costs – i.e. the costs of doing business; property damage including vehicle and equipment loss and repair. Average economic cost per death, injury or crash, 2004:

Death	$1,130,000
Non-fatal disabling injury	$ 49,700
Property damage / other non-disabling injury	$ 7,400

In addition to the direct economic cost components relating to a motor vehicle accident, a 'comprehensive cost' is also determined. This includes a measure of the value of lost quality of life, this has been obtained through empirical studies of what people *actually pay* to reduce their safety and health risks. These cost values do not represent real income or expenses incurred, therefore they can only be used for cost-benefit analysis calculations.

Death	$3,760,000
Incapacitating injury	$ 188,000
Non-incapacitating injury	$ 48,200
Possible injury	$ 22,900
No injury	$ 2,100 [ibid.]

The same source goes on to discuss work injury costs, but cites that the values are partial direct costs only, not including property or equipment damage or loss, and not including the researched values of what people actually pay to reduce their personal risk profiles.

Work death	$1,150,000
Work disabling injury	$ 34,000 [ibid.]

In the UK, the government Department for Transport (DfT) has done this type of calculation for assessing the economic viability of road improvement schemes. The values have been adopted widely in UK industry, although this near blind acceptance is open to debate, and a number of modifications are used.

Year 2004 figures from the DfT (latest available at time of writing) are £1.384m per prevented fatality [DfT 2005]. The values currently in use for preventing serious and slight injuries are £155,560 and £11,990 respectively (at June 2004 prices). It is worth noting that these do correspond to differences of roughly one order of magnitude and so do serve well in logarithmic scale risk analysis matrices. The total number of all recorded UK road accidents in 2004 was 2,978 fatal, 207,410 injury accidents and 3.1m damage only. The total value of prevention of all road accidents in 2004 has been estimated to be £18,004 million [ibid.].

The DfT also produce prevention values 'per fatal accident' as well as 'per individual fatality' – note the change in units. The values currently in use for preventing accidents are £1,573,000 for a fatal accident, £184,200 for a serious injury accident, £18,500 for a slight injury accident and £1,650 for a damage only accident.

The calculated values relate to the total value to the community of the benefits of preventing road traffic accidents. The UK cost includes direct factors such as lost economic output, medical and emergency service costs. It also includes a significant contribution from 'human costs'. This is a similar factor to the US value of lost quality of life, but it is calculated in a completely different way. In the UK, this factor is based on what people said they were 'willing to pay' annually to avoid the accident event. Techniques to establish monetary values for this type of value impact generally involve the inference of a price, through either a revealed preference or stated preference approach. Revealed preference techniques involve inferring an implicit price revealed indirectly by examining consumer's behaviour in a similar or related situation.

The 2004 figures for the construct of the DfT value for the prevention of a fatal accident is given as shown in Table 12.1 [ibid.].

Table 12.1 Construct of Value for the Prevention of a Fatal Accident

Element of Cost	Value (£)
Lost Output	522,639
Medical Costs	5,469
Human Costs	1,033,783
Police Costs	1,607
Insurance and Admin	254
Property Damage	9,465
Total	1,573,217

As noted above this figure has been adopted by many industries in the UK, but with the addition of modification factors. The Health and Safety Executive has published guidance on the use of the DfT value in ALARP cost-benefit analyses [HSE 2004].

Le Guen, Hallett and Golob produced a paper in 2003 on the "Value of preventing a Fatality" which was circulated to HSE Board members and presented to the Risk Assessment Liaison Group (RALG) (RALG/Sep00/03). That paper discussed the ratio of the cost of preventing a fatality (CPF) to the value of preventing a fatality (VPF). The starting point for VPF was taken to be the DETR [now known as DfT] figure of approx. £1m [now £1.5m] used in new road schemes. Other values of VPF were then described. These were 2 x DETR for deaths from cancer and 3 x DETR for some aspects of railway safety.

These multiplication factors are based on the public perception of the horror or dread factor of dying in a particular way – see chapter 7. It may not be totally correct to multiply the whole figure of all components in this way – as the factor is based on public perception, perhaps only the human cost part should be effected.

In recent research, this concept was pursued for the military domain [Maguire 2005], where the straight use of the DfT figure was questioned, as the component factors did not always transfer in an equivalent way. The three main contributory components were assessed, output loss, human costs and property damage.

When considering a comparison between civilian and military employees it is difficult to argue a large difference. A limited review of the military salaries would appear to justify that the military has financial benefits comparable with civilian society. However, two factors need to be considered – occupation and age. All military fatalities happen to employed persons, this would appear to justify an increase in this value, they are all earning so have more to not-contribute. However, all the military fatalities happened to persons of employable age i.e. not children (the largest group of civilians involved in road traffic accidents). This means that military persons have already had some employer contributions made and so this would justify a decrease in the remaining to be paid. It is recommended that these values be allowed to cancel each other out.

For the human cost component … It may be argued that service personnel have accepted the risk of their profession. The public perception of a fatality of a member of the services may not be seen with the same 'dread factor' as a cancer fatality – or even as a road traffic fatality. … Whilst it is absolutely true that an accidental military fatality is catastrophic, especially for the relatives, friends and colleagues – the public perception of the incident is not likely to be as severe as that for a child being run over and killed. Pending further in-depth research, it is recommended that the human cost element be reduced by a factor of 50%.

During 2004, seven of the UK military fatal accidents involved loss or serious damage (category 4 or 5) to MoD aircraft, both fixed and rotary wing. These incidents resulted in the loss of 9 aircraft. For the military fatalities in 2004 the total property damage is estimated to be at least £225.5m. This value, averaged out (as the element is done for the civilian calculation) over the number of fatal accidents, will give the property damage element for the military domain … The property loss element for military accidents should be considered to be at the £7.5m mark, base-lined to financial year 2004.

This calculation gives a justified military fatal accident cost of somewhere around the £8.5m mark, which is far in excess of the UK DfT figure. This could have dramatic consequences for cost-benefit analysis used in ALARP justifications, and hence in the content of safety cases in this and many other industries. It may be time to review the value of a prevented fatality in many industries if safety engineering and safety cases are to continue to be taken seriously.

Cost Indicators from Criminal Fines

There have been three notable criminal investigations in the UK and US where record industry fines have been imposed following fatal accidents. From these values, it is possible to evaluate the cost impact of a fatality – this is not a method for working out the 'value of a life', it is just an average value of the fine imposed per fatality.

The three incidents are the rail disaster at Hatfield in England, the Larkhall gas explosion in Scotland and the Texas City refinery explosion. In these three events four rail travellers were killed at Hatfield, a family of four was killed at Larkhall and 15 workers were killed in Texas City, making a total of 23 people. Following the investigations the fines imposed were £13m, £15m and US$21m, a total of around £42m or US$60m.

The mathematics are trivial - £1.83m per fatality (= US$2.61m per fatality). These values are consistent with the values calculated by the respective governments for the costs of preventing a fatality.

Cost Indicators from Other Fines

Not all fines are this high, as the level does depend on the individual event circumstances, where the risk was being controlled, the level of acceptance of the risk by the people impacted, and how good the safety analyses were of the organisations involved. The other major factor that determines a level of fine is wilful intent. In the US, the Department of Labor has published a table of violation categories and possible penalties, where a wilful violation that results in a fatality has a maximum individual fine of US$250,000, and/or six months in prison [OSHA 2003]. This is the value per violation, with no limit on the number of concurrent violations. The full table is as shown in Table 12.2 below.

Table 12.2 OSHA's Violation Categories and Possible Penalties

Type of violation	Minimum penalty per violation	Maximum penalty per violation
Other than serious		$7,000
Serious	$100	$7,000
Wilful	$5,000	$70,000
Wilful with fatality, first conviction		$250,000 (Person) $500,000 (Company) and/or 6 months in prison

Type of violation	Minimum penalty per violation	Maximum penalty per violation
Wilful with fatality, second conviction		$250,000 (Person) $500,000 (Company) and/or 12 months in prison
Repeated	$5,000	$70,000
Failure to abate		$7,000 per day

Notes

DfT 2005, "Highways economic note No.1 : 2004", Department for Transport, London, 2005.

HSE 2002a, "Costing study of incidents at a NHS hospital", F ORG/C/S/12, The Health and Safety Executive, Bootle 2002.

HSE 2002b, "Costing study of incidents on a large construction site", FORG/C/S 8, The Health and Safety Executive, Bootle 2002.

HSE 2002c, "Costing study of incidents at a cheque clearing company", FORG/C/S 7, The Health and Safety Executive, Bootle 2002.

HSE 2002d, "Costing study of incidents at an oil production platform", FORG/C/S 9, The Health and Safety Executive, Bootle 2002.

HSE 2002e, "Costing study of incidents at a transport company", FORG/C/S 11, The Health and Safety Executive, Bootle 2002.

Maguire 2005, "Just what is the value of a prevented fatality in the Services?", Equipment Safety Assurance Symposium, MoD Abbeywood, Bristol, November, 2005.

NSC 2005, "Estimating the costs of unintentional injuries, 2004", National Safety Council, Itasca, USA, December 2005.

OSHA 2003, "Violation Categories and Possible Penalties", OSHA 2056-07R 2003.

Chapter Thirteen

Techniques and Tools for Safety Cases

Introduction

This section introduces a description and discussion of a number of tools used by the safety team in developing the safety case evidence and constructs. The types of tools and techniques used include HAZOPs, Fault-trees, Event-trees, Zonal Analysis, FMEAs, SWIFT and Human Hazard Analysis. Examples will be brought out along with some demonstration of the use of graphical output from typical software tools.

Many of the tools and techniques are utilised for both of the main tasks in safety case development – hazard identification and hazard
. The first seeks to identify all reasonable hazards related to the system or equipment of interest, the second attempts to evaluate (qualitatively or quantitatively), organise and even prioritise the hazards.

HAZOP

A HAZOP is a HAZard and OPerability study, it is a technique for systematic examination of any system to assess the hazard potential due to incorrect operation of the component parts of the system. It can be further developed to analyse the consequential effects of the occurrence of the hazard on the whole system.

It originated in the chemical industry, but is now used extensively in many other industry areas. Originally, the needed information items to carry out a HAZOP were process flow diagrams (PFDs) and piping and instrumentation diagrams (P&IDs), i.e. the design and construction of the system. A team of engineers and specialists were also needed covering design, operations and maintenance. The method required the system to be divided up into component parts (originally called 'nodes'), followed by a series of guidewords applied to each node to assess what would happen if the guideword did occur. To enable efficient use of resource a standard worksheet was produced with cells and headings for each answer.

To carry out a HAZOP, these resources will need to be available; the system design, the team of experts, the applicable guidewords and a method of recording the information.

The guidewords used in the chemical industry are given in table 13.1 along with their chemical-based interpretations. For a non-chemical system, the guidewords will probably have to change. This is fine, as long as a record of the actual guidewords used and the rationale for their use is kept.

On completion of the HAZOP, an initial report is usually issued containing the recommended actions to be applied to the system. A further, final report is issued when all the actions have been completed and this becomes the audit trail of the study.

Table 13.1 HAZOP Guideword Descriptions [Wong 2002]

Guide Word	Typical deviation	Possible explanation
No, None	No flow	Diverted, blockage, closed valve
More	Flow	More pumps, inward leaks
	Pressure	Excess flow, blockage, closed valve
	Temperature	Cooling failure
Less	Flow	Blocked suction, drain with closed vent
	Pressure	[Under flow]
	Temperature	[Excess cooling]
As well as	Contamination	Carry over, inward leaks at valves
Part of	Composition	Incorrect composition
Other than	Abnormal situations	Failure of services
	Maintenance	Isolation, venting, draining
	Abnormal operations	Start-up, part-load

The strengths of HAZOP are that it is widely understood, it uses the experience of operators as well system designers, it works well with both technical faults and human-based errors and it allows recognition of existing safeguards [DNV 2001]. Its weaknesses are that it strongly depends on the facilitation of the team leader, it has been optimised for chemical process hazards and needs modification for other areas and the full recording documentation is resource intensive to produce [ibid.].

Structured What-if Technique (SWIFT)

SWIFT is the Structured What-If Technique for Hazard Identification that was developed as an efficient alternative to the Hazard Operability Studies (HAZOP). Whereas the HAZOP examines a system item-by-item, procedure-by-procedure, the SWIFT is a systems oriented technique that examines complete systems or sub-systems. The technique utilises a structured brainstorming approach by a team of experienced experts augmented by supplemental questions from a checklist.

The "What-If" questions, which can be posed by any team member (including the team leader and recorder), are structured according to various hazard categories. When the team is no longer able to identify additional questions in a category, a category specific checklist is consulted to help prompt additional ideas and ensure completeness.

A suitable taxonomy may be as follows:

1) Initially the design intent, the conditions prevailing and the boundary are agreed for the topic under discussion. Drawings of the topic are also made available for discussion and reference.
2) The discussion is commenced through a review of the circumstance of the system and the regulatory framework in place. The discussion should cover;
   a) Hazards of the activity or procedure
   b) Previous incidents
   c) Engineering and administrative controls
   d) Siting/layout issues
   e) Qualitative evaluation of safety and health effects
   f) Any other regulatory issues.
3) Discussion is then initiated using question categories relating to the topic under review. The structure of questioning in the original SWIFT, which was developed for the process industry, is provided by the following categories:
   a) Material problems
   b) External effects or influences
   c) Operating errors and other human factors
   d) Analytical or sampling errors
   e) Equipment or instrumentation malfunction
   f) Process upsets of unspecified origin
   g) Utility failures
   h) Integrity failure or loss of containment
   i) Emergency operations
   j) Environmental release.
4) The asking of the 'what-if?' questions follows. The intent is to ask questions that will cause the group to carefully consider and think through the potential scenarios and ultimate consequences that such an error or failure might lead to.
5) The possible consequences are then examined and if the team considers current detection/safeguards or mitigation to be sufficient, the next question is raised. The recorder should enter a brief summary of the discussion in the SWIFT logsheet.
6) If the team is not satisfied with the level of protection or perceives a need for further analysis, recommendations for further action are proposed and recorded. Such recommendations shall include a brief description of the potential hazard, a description of what equipment, instrumentation or procedures currently in place are relied upon to prevent the development of the hazard and finally, the objectives, which must be achieved to provide a solution to the potential problem.

The strengths of SWIFT are that it is very flexible and applicable to almost any equipment or system of interest at any stage of the lifecycle. It uses the experience of operators and it is quick to utilise as it avoids unnecessary repetition of illogical guidewords. Its weaknesses are that as it works at the system level, component hazards may be overlooked; preparation time is significant as the what-if checklist prompts do need to be developed at the start. Finally, the benefit from the technique depends strongly on the experience of the leader and knowledge depth of the team members [DNV 2001].

Fault-tree Analysis

Fault-tree analysis uses logic diagrams that display the interrelationships between a potential critical occurrence in a system and the contributing events for it. The reasons may be software induced, human induced, environmental or operational conditions, or any combination of the above. The fault-tree diagram illustrates the operational states of the system's components (via basic events) and the connections between these basic events and the whole system state (the top event). Graphical symbols are used to represent the connections, these are called logic gates. The propagation of an event sequence through a logic gate is determined by the inputs to it.

To start a fault-tree, the undesired/accident/failure top event has to be identified. From this point, all the possible events that could make a contribution to the failure have to be identified. This process is repeated for each sub-event and each sub-sub-event until all the basic events have been identified to the resolution level required. The basic events ought to be fundamental engineering components or human tasks where it is possible to make defensible arguments about the likelihood of failure. The logic rules of the tree structure allow the basic failure rates to be mathematically processed, so that the overall probability of the top event/failure can be calculated.

A fault-tree analysis is normally carried out in five steps [Høyland & Rausand 1994];

Definition of the problem (top event) and system boundary. A description of 'what', 'when' and 'where' has to be made. The system boundary does not only contain the physical boundary, but also the operational state of the system, any external stresses on the system and the level of resolution of the analysis.

Construction of the fault-tree. When the top event has been satisfactorily described the top structure needs to be completed. The faults that are the immediate, necessary and sufficient to cause the top events are to be established (perhaps via related FMECA analysis). The analysis is deductive and is carried out, one layer at a time, by repeatedly asking "What are the reasons for this event?". Care has to be taken when labelling events, as unintended duplicate events can be introduced leading to difficult and incorrect analysis later.

Identification of cut sets (or path sets). A cut set is a set of basic events whose occurrence ensures that the top event will take place. These are valuable information items for subsequent analysis.

Qualitative analysis of the fault-tree and cut sets. This evaluation may be carried out on the basis of the cut set identification information. Cut sets with lower numbers (or orders) of basic events are more critical than those with a higher number. Another factor in this analysis is the type of basic event, these may be classified and ranked in order as; human error, active system errors and passive system errors.

Quantitative analysis of the fault-tree and cut sets. This analysis may be carried out on the basis of basic event occurrence probabilities, where these can be reasonably determined. Approximation or exact system reliability calculations may be possible depending on the level of independence between basic events. For smaller Fault-trees the quantification task is easier, as trees become larger, the quantification effort increases almost exponentially.

Other methodologies are certainly available – there does not appear to be a universal 'standard' way of producing a fault-tree diagram, although the logic gate design is consistent. It seems that slightly different taxonomies have developed for different industries – computing, chemistry and defence. Essentially though, the fundamentals have remained as above, and you should end up with a tree structure as shown in Figure 13.1 [Maguire 2005]. In this example, Gates 1 and 3 are 'AND' gates, such that a positive output requires both inputs to be positive. Gate 2 is an 'OR' gate, such that a positive output requires a positive input to input 1 OR input 2.

The use of mathematical logic and graphical representation has led to this technique becoming one of the first to be developed for use on a computer. Numerous tools now exist across industry, you can take your pick. Depending on which standard you are obliged to follow, how much basic event data you want to have included in databases and how many logic-based analysis functions you want to employ, you can almost certainly find a product to match your needs.

The strengths of fault tree analysis are that it is used and understood widely, it is good at representing many types of hazard and it is often the only technique that can generate a credible likelihood for complex systems [DNV 2001].

There are some traps for the unaware when using these computer based tools. Firstly, as with many computer-based tools, the functionality packed into Fault-tree tools can be far above the actual requirements of the system analysis required for safety case work. Authors and readers may not easily understand some of the complex logic-based analysis available, and if this cannot be appreciated, how can it be of value or used with competence?

It is assumed that all events are mutually independent, which can cause problems for parallel-based systems, and it rapidly loses clarity when applied to systems that do not have '0' and '1' indices for operational modes, i.e. weather hazards. It is also assumed that all the units are compatible, so great care must be taken when combining basic events, otherwise you can end up with illogical and completely nonsense units e.g. 'Failures per hour^2 per demand'

Also, the computer application always does exactly what it is ordered to do. This is perfectly fine, but when the user has a different mental model of the logic structure compared to the computer's operational model, real problems ensue. Each of the basic events needs to have a label, and these labels can be repeated in the same Fault-tree only when it is the intention to include exactly the same event. I have reviewed fault-trees where there were tens of basic events all labelled as 'Pilot error'. Of course they were not the same pilot error, but the computer didn't know that. The quantification and calculation of the probability of the top event was ridiculously invalid.

The final problem with Fault-trees is that they can become unmanageably large, running over tens and hundreds of pages. Navigation through the computer screen or in paper format becomes near impossible – almost to the point that the tool becomes useless, and therefore it is no longer a tool at all!

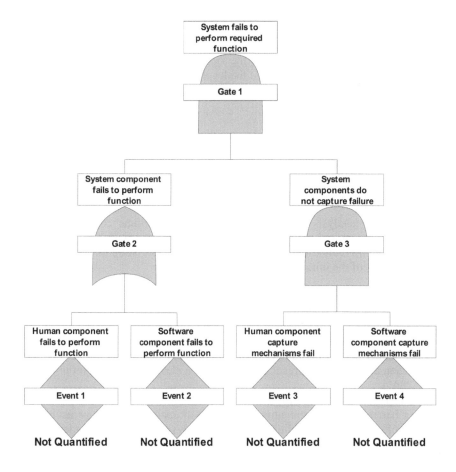

Figure 13.1 Example of a Fault-tree

Event-tree Analysis

An event-tree is a logic diagram that displays the potential consequences that could arise from a critical initiating event in a system. The event may be induced by software, hardware, human, environmental or operational conditions, or any combination of the above. The event-tree portrays all credible system operating permutations and traces each path from initiation to eventual success or failure.

To start an event-tree, the undesired initiating event has to be identified. From this point, all the possible paths that could contribute to a hazardous event have to be identified. In this process a number of points are identified that determine whether the initiating event will propagate to a safe or a potentially unsafe condition. These points may be considered as barriers.

An event-tree can be carried out in the following steps;

1. Identification of a system boundary and within it the barriers that are designed to deal with the accidental event. The system boundary does not only contain the physical boundary, but also the operational state of the system, any external stresses on the system and the level of resolution of the analysis.
2. Identification and definition of an initiating event that may give rise to unwanted consequences.
3. Construction of the event-tree.
4. Identification and description of the potential resulting accident sequences.
5. Qualitative identification of the barrier functions; *or*
6. Quantitative analysis of the event-tree through determining the frequency of the initiating event and the (conditional) probabilities of the branches of the Event tree and calculation of the probabilities/frequencies of the identified outcomes; *or*
7. Quantitative analysis of the event-tree to determine the conditional probabilities required for each branch in order to ensure that the probabilities for the identified undesirable outcomes are acceptable, based on the estimated probabilities of the initiating events.

The strengths of event tree analysis are that it is widely used and understood, it is suitable for many hazard types and it has a clear and logical format of presentation. However, it is not efficient where there are many events in an accident sequence as this can result in many redundant branches; it can lose clarity when applied to features that do not have a clear binary function in failure [DNV 2001].

Event trees enable one to identify the functionality within systems that provides barriers that prevent unwanted events propagating of cause accidents. There is often a required additional 'barrier analysis' step that considers these hardware, software or human barriers in greater depth. This assesses whether there is potential for undetected or common-mode failures in the barriers themselves that might lead to them having less effectiveness than that included or required by the event-tree analysis. For example the initial unwanted event might be a power failure at a critical time, but this might also disable the locks or alarms that are designed to prevent the event leading to an accident.

Zonal Analysis

Zonal analysis is an assessment of the physical instantiation of the system and its component parts in the working environment and operating domain. It is used to determine the consequences of effects of interactions with interfacing and co-located systems. It is also used to identify sources of common cause failures, e.g. flooding and service disruption, and for this reason it is sometimes known as 'Common cause analysis'. Transportation and storage effects can be considered, as the analysis should look at all operating domains of the system, even temporary ones [MoD 1996].

The analysis should be based on the latest available design concepts and if possible a first representation of the system in its operating environment. The analysis should consider matters such as:

- Moving parts clearance
- Temperature effects – heating and cooling
- Corners, edges, and sides
- Positioning and access to the system for installation and removal
- Provision and back-up of power supplies
- Environment, weather and water effects
- Neighbouring equipment analysis
- Spatial relationship between operator and equipment.

The strength of zonal analysis is that it looks at the system as a whole in its environment. It can be facilitated through an audit-type procedure if an appropriate checklist is developed and approved. It is also important for starting to bring in analysis of the human operators. Its weaknesses are that it does analyse at what is considered the highest level, component level hazards are not well captured. There are no published standard formats for the analysis, so the integrity of recording and reporting is very much down to the capability of the analysts.

Failure Mode Effect Analysis (FMEA)

A Failure Mode and Effects Analysis (FMEA) is an inductive bottom-up method of analysing system designs or development processes in order to evaluate the potential for failures. It consists of defining what can fail and the way it can fail (failure modes) and determining the effect of each failure mode on the system.

An important extension of FMEA also includes analysing criticality, which is how severe the effects of a failure mode are on system operation. When criticality is considered in FMEA, the name is changed to Failure Mode, Effects, and Criticality Analysis (FMECA).

While performing FMEA or FMECA, the failure modes and causes of the failures are examined. This analysis has been used to understand the product responses to the failure. Steps may then be suggested to change the design or process to eliminate the failure, reduce its impact, or compensate for the failure should it occur.

There are several different standards available for FMECAs. Some examples are the Military Standard (Mil-Std-1629A), the Society of Automotive Engineers Ground Vehicle Recommended Practice (SAE J1739), and the Aerospace Recommended Practice (ARP5580). All three of the above standards provide general FMEA forms and documents, identify criteria for the quantification of risk associated with potential failures, and offer general guidelines on the mechanics of completing FMEAs. Many people use a combination of all these different standards, modifying them to suit their needs for their particular applications [Høyland & Rausand 1994].

The analysis uses a dedicated form that records failure mode data, it would typically include [DNV 2001]:

- Component name and reference
- Function of the component
- Possible failure modes [an entry for each mode]
- Cause of failure
- How the failure would be detected
- Effects of failure on primary [local] system function
- Effects of failure on whole system function
- Necessary preventative/repair action
- Rating of frequency of failure
- Rating of overall severity of failure.

Because of the nature of the data to be recorded and the potential extensive nature of information needed to be collected, this technique has generated many computer based tools. Again, depending on exactly what sort of system you want to analyse, which standard you are following and how much pre-loaded failure rate data you require, there will most probably be a tool that you can purchase. Alternatively, a simple computer tool that accepts tables can be sufficient for small scale use.

The strengths of FMEA/FMECA are that is it widely used and understood, it can be carried out by a single analyst, it is systematic and comprehensive and it can identify safety-critical areas where single failures could lead to whole system failure. Its weaknesses are that the benefit obtained depends on the experience of the analyst, it requires a hierarchical system structure representation, which has to be developed if it is not already available, and it has particular difficulty in handling multiple failures and failures with a human origin [ibid.].

Human Hazard Analysis

HHA is an analytical approach to the identification of potential errors and their likely mechanisms. The HHA methodology uses an error taxonomy applied to a task analysis to elucidate potential errors. It has been principally designed for use in the aircraft maintenance domain, but may also be of use where humans follow a strict procedural regime. The use of a taxonomy to identify error mechanisms makes identification of errors possible through analysis rather than relying solely on memory and existing knowledge (referred to as experience and engineering judgement) associated with the design characteristics of the component or system itself. HHA provides a proactive approach in which the potential for error can be identified through the knowledge of how errors are produced without having to happen first [JRC 2003].

The strengths of this approach are to make the error identification process more consistent both across and within users, to widen the scope of the potential errors considered, to provide traceability of HF issues considered during the design process, and to store the analyses for future reference. The weaknesses are that this process requires considerable maintenance and maintainability expertise to be performed effectively. However, relating to certain stages of the HHA, prompts have been developed to help individuals with less maintenance expertise. This can help make less knowledgeable individuals perform more thorough analyses, but should also be of assistance to expert users to minimise day-to-day fluctuations in what is salient to the user. A dedicated computer-based tool has been developed by the European Commission Joint Research Centre.

Human Reliability Analysis

Human reliability analysis (HRA) has the purpose of identifying the probability of different types of human error during system operation, to form the basis for system modification and other hazard reduction measures [MoD 2004].

Several techniques are available for HRA, all of which require task synthesis at the primary input. The quantification of the error rate is carried out by generating empirical data and using a fault-tree method, or by the use of principles based on actuarial data. This is data generated by recording and combining error rates from other domains and satisfying yourself that the values are representative to your particular system. Several tools are available to assist and guide users, all have rather catchy acronyms (it seems to be a requirement to have one!) including;

THERP   Technique for Human Error Rate Prediction
HEART   Human Error Assessment and Reduction Technique
SLIM   Success Likelihood Index Methodology
THETA   Top-down Human Error and Task Analysis.

The HEART method involves classification of each of the tasks identified in a task analysis exercise. The next step is to assign a nominal human-error probability from actuarial data to the task according to its classification. The data has been gathered from multiple researchers assessing various industries including nuclear control rooms, the bridges of ships, aircraft operations and other command and control domains. The next stage is to identify any obvious error-producing conditions and note the likelihood multiplication factor for each condition. Finally, there is a calculation using the assigned values to produce a final estimated human error probability rate. Table 13.2 shows an extract from the potential error-producing conditions with the multiplication factors, and Table 13.3 shows the task classification options. In both cases the quantified values are given – the nominal error probability and the multiplication factor for the error producing conditions [Reason 1997].

Table 13.2 HEART Error Producing Conditions (extract)

Error producing condition	Multiplication factor
Unfamiliarity with a situation that is potentially important, but which is novel or occurs only infrequently	x17
Shortage of time for error detection and correction	x11
"Suppression of feature information" that is too accessible	x9
Absence or poverty of spatial and functional information	x8
No obvious means of reversing an unintended action	x8
Channel capacity overload, particularly one caused by the simultaneous presentation of non-redundant information	x6
The need to unlearn a technique and apply one that required opposing philosophy	x6
Mis-match between real and perceived task	x4
No clear or timely confirmation that an action has had an effect on the intended portion of the system	x3
Operator inexperience (newly qualified operator)	x3
Unreliable instrumentation that is recognised as such	x1.6
No obvious way to keep track of progress during task	x1.4
High level of emotional stress	x1.3
Low workforce morale	x1.2

Table 13.3 HEART Task Classifications

	Generic task type	Nominal error probabilities (5th-95th percentile bounds)
A	Totally unfamiliar, performed at speed with no idea of likely consequences	0.55 (0.35 – 0.97
B	Shift or restore a system to a new or the original state on a single attempt without supervision or procedures	0.26 (0.14 – 0.42)
C	Complex task requiring a high level of	0.16

Generic task type		Nominal error probabilities (5th-95th percentile bounds)
	comprehension and skill	(0.12 – 0.28)
D	Fairly simple task performed rapidly or given scant attention	0.09 (0.06 – 0.13)
E	Routine, highly practised, rapid task involving a relatively low level of skill	0.02 (0.007 – 0.045)
F	Restore or shift a system to the original or new state following procedures, with some checking	0.003 (0.0008 – 0.007)
G	Completely familiar, well designed, highly practised routine task, often repeated and performed by well motivated, highly trained individuals with time to correct failures but without significant job aids	0.0004 (0.00008 – 0.009)
H	Respond correctly to system even when there is an augmented or automated supervisory system providing accurate interpretation of the system state	0.00002 (0.000006 – 0.00009)

Stored Energy Analysis

This analysis technique primarily considers the release of stored energy, as hypothetically, it can take around 20 joules to kill someone. When a human is exposed to a rapid energy release injury or death is possible. The pre-requisites are a design and use description, preferably with some detail of the operational environment, and a developed checklist of energy types. The system would then be assessed against each energy type on the checklist to see if it had any of that energy type. The existing controls of that energy source are then listed and assessed using engineering judgement, if the controls are not deemed sufficient when compared to the quantity of energy available, further mitigation and control is recommended. A typical checklist would look something like this:

- Electrical energy
- Explosive energy
- Combustion energy
- Thermal energy
- Cryogenic energy
- Pressurised system energy
- Vacuum energy
- Kinetic energy
- Potential energy
- Buoyancy energy
- Laser energy
- Ionising and non-ionising radiation energy

- Noise and vibration energy
- Electro-magnetic energy.

The strengths of this type of analysis are that it is very easy to understand and quantify – equations exist for calculating potential energy, radiation energy etc. The assessment can be done by a single analyst and can easily be duplicated for audit purposes. The weaknesses of the technique are that it does not consider all hazardous sources e.g. toxicity, and at present it is only used for human impact, although development for other impact domains is possible.

Summary

There are many and varied tools and techniques for carrying out hazard identification and hazard analysis, I'm sorry if I haven't summarised your particular favourite. My main recommendation for taking part in hazard identification or analysis is not to use just one single tool or technique, none of them give complete and rigorous coverage over all hazard types to the correct level of fidelity. Reliance on a single method will leave you open to criticism, and at best your safety case will have to be re-worked.

The final piece of advice to leave you with at the end of the chapter of tools and techniques is, irrespective of what you actually do, please, please record it. These records will become high quality evidence for a safety case, and you are unlikely to be able to afford the time, trouble and money (sound familiar?) to go and repeat the work.

Notes

NDV 2001, "Marine Risk Assessment", Offshore technology report 2001/063, Section 2. Det Norske Veritas, London, 2001.

Høyland & Rausand 1994, "System Reliability Theory – Models and statistical methods" Section 3.3, John Wiley & Sons, Inc, New York, 1994.

JRC 2003, "Annual Report 2003", European Commission, Joint Research Centre, Human Factors Sector, Ispra, Italy, 2003.

Maguire 2005, "Human Factors and Software Factors – A powerfully reliable system", IMechE symposium on the economics of reliability, London, 2005.

MoD 2004, "Human Factors for Designers of Systems, Part 15: Principles and Processes", Ministry of Defence Standard 00-25, MoD, London July 2004.

Reason 1997, "Managing the Risks of Organizational Accidents", p142-145 Ashgate Publishing, Aldershot, 1997.

Chapter Fourteen

The Hazard Log

The Role of the Hazard Log

The hazard log is a store of information about the hazards relevant to the system or product in question. It should act as a central control system for the management and demonstration of on-going safety activities. It should contain the documentary evidence on the handling of risks that will be vital to the construction of a safety case.

The hazard log should contain information on all possible hazards, even ones that are considered to be incredible. One of the worst comments that can be made about an accident or dangerous event is " ... we hadn't even thought that could happen". Even when a hazard has been analysed and identified as tolerable, the information about the hazard should be retained as a historical record of progress. The choice of information captured is an open one, but particular structures are recommended according to the industry and system type that is being considered. However, a number of key information items are consistent across them.

The Requirement for a Hazard Log

Many of the standards and regulations cited in earlier chapters of this book have the requirement to keep a record of the hazards and risks that are considered when conducting a safety review – even if an explicit 'safety case' is not being developed. And as with the precise name used for the safety study, the precise name used for the record log is also different.

The name used in the UK is pretty consistent as the 'hazard log', the defence industry standard for safety management of defence systems [MoD 2004], has the following requirement at section 10.4;

> The contractor shall implement a Hazard Log and manage it as part of the Safety Management System. The Hazard Log shall be the primary mechanism for providing traceability of the risk management process and assurance of the effecting management of hazards and accidents. The Hazard Log shall be updated through the life of the contract to ensure that it accurately reflects the risk management activity.

The rail industry has a very similar concept. The year 2000 edition of the 'Yellow Book' or 'Engineering Safety Management' has the following guidance at section 13.1.2 relating to the hazard log [Railtrack 2000];

You should keep records to show that you have followed the safety plan. These records may include the results of design activity, analyses, tests, reviews and meetings. You should keep a hazard log which records all the possible hazards identified and describes the action taken to get rid of them, or reduce their likelihood or severity to an acceptable level.

In the US the phrase 'hazard log' is certainly known and used, but not that widely. Here it is used in the software environment [Leveson 2003];

All hazards in the system must be entered into a hazard log. A hazard log is essential to any safety effort. The hazard log, part of the safety information system, tracks information about hazards from their initial identification through elimination or control.

It may also be known as the 'Hazard Tracking Log' [FAA 2000] as part of a hazard tracking and risk resolution process. This FAA handbook offers an example tracking log and then states that;

All identified hazards should be tracked until closed out. This occurs when the hazard controls have been validated and verified. Validation is the consideration of the effectiveness and applicability of a control, verification ... is the act of confirming that the control has been formally implemented.

Also, many industries in Europe have acknowledged the requirement and value of utilising the hazard log concept [Schäbe 2002].

The hazard analyses are focussed on the systematic identification of hazards and reduction of risks. This process is started with the Preliminary Hazard Analysis and carried on with the help of the Hazard Log. The Hazard Log is the central document that documents all identified hazards. It is used to trace the process of hazard reduction.

It appears, and I concur, that the hazard log is an integral and essential component part of the safety case. Whilst the references are not all mandatory standards, the recurring theme and description of the purpose of the hazard log, does give credibility and force to the position and value of recording hazard information and the progress over time of the work done to remove or reduce the hazard.

The intensive nature of the recording process and the on-going safety value of the evidence has led to the development of a number of software-based tools for recording all the data. Essentially though, any table supporting software will do for a simple hazard log, all that is needed is the correct headings and data fields. However, in high risk, high value contracts, it is often essential to have permanent records of the hazard log and reduction process. Certain software products will keep a permanent memory that is read only, so that historical data can not be modified without a record being made. This keeps accurate configuration control over the database, and is potentially strong evidence for independent audit and legal reference.

The Content of a Hazard Log

A simple hazard log need only record a few basic date items, such as a unique description, remarks and mitigations. For a more sophisticated hazard log, with therefore more utility, significantly more information should be recorded. An older, withdrawn UK defence standard from 1996 gives guidance on the content of a hazard log – this guidance has been removed from the 2004 update, perhaps indicating that this information is becoming more widely known.

This older standard indicates five parts to a hazard log, these are as follows [MoD 1996];

1. Introduction. This part should describe the system under analysis and purpose of the hazard log. It should indicate the environmental and safety criteria to which the system safety is being assessed.

2. Accident data. This part should give sufficient information to identify the accident sequence linking each accident with the hazards and events that cause it. It should also present the severity and probability rating for each accident according to the developed risk matrix.

3. Hazard data. This part should give sufficient information to identify each hazard and the risk reduction process applicable to each. A textural summary of the contemporary status of the hazard should be included.

4. Statement of risk classification. This part should be a brief statement of the contemporary whole system risk classification based on the status of all the hazards. It should contain enough information to be a stand-alone piece of evidence for the safety case.

5. The journal. This part should be constructed to provide a continuous historical record of the compilation and development of the hazard log. It should refer out to any appropriate risk analysis, tests and evaluations and the minutes of any project safety meetings – particularly those meetings where the hazard log was discussed and endorsed.

The FAA hazard tracking and risk resolution log introduced above also contains five sections, but in a much simpler table format. Its headings are item/function, operating phase, hazard description, controls and corrective action & status. It is suggested that the log could also serve as an analysis tool in its own right, but it is also recognised that it does not meet the normal (US) definition of a hazard analysis, as it does not include an analysis of severity or probability levels.

Examples of Real Hazard Logs

In this section I am going to show a few examples of hazard logs of different types. It is not for me to judge them, as I do not know the full facts behind each.

The first example is from a presentation made to a project safety meeting with a request for endorsement. The meeting was a special co-ordination meeting for the Bay of Bengal Air Traffic Flow Management Operational Trial (BOBCAT) [ICAO 2006]. The paper presented a draft safety assessment (safety case?) and hazard log for the operational trial of the BOBCAT system for review, update and acceptance. The extract is just the first hazard, but it should be a standard one for nearly any system involving the use and exchange of data.

Table 14.1 Example of a Simple Hazard Log

	Hazard 1
Description	Non-standard, incorrect or corrupt data leading to erroneous advisory information
Remarks	The BOBCAT is a computerised system with user access via the public internet. This hazard identifies the possibility of incorrect data being presented to or utilized by BOBCAT, resulting in erroneous advisory information being promulgated by BOBCAT.
Mitigation	1) BOBCAT provides advisory information only, ATS providers retain responsibility for tactical ATS and traffic management. 2) Hardware – The BOBCAT 'concept of operations' includes details of system hardware architecture which incorporates contemporary firewall protection to ensure no unauthorized access is obtained, in particular to application and database servers. 3) Software – incorporates checking algorithms to ensure aircraft cannot be scheduled at the same gateway fix at the same time and at the same flight level. 4) AEROTHAI support engineers will monitor the BOBCAT error log and provide support for ATFMU staff to resolve non-standard, incorrect or corrupt data issues. 5) Paper trial – paper trials/simulations of the full functionality of BOBCAT prior to operational trail comprises validation exercises in order to identify data errors and other issues.

As was noted above, this is a fairly simple log of one hazard – true it does contain significant technical detail, but it does not contain time and responsibility data, nor does it show a development schedule over time. It probably doesn't actually need these things for the purpose it has for the meeting. One of the interesting points to bring out is the number of mitigations in place. The first dictates that the system is actually only advisory and so implies that it cannot be relied on as the sole source of this critical flight information. The second and third are mitigations based on the properties of the product itself and the protection that is has. The fourth is based on the on-going management of any incorrect or corruption issues and the last one indicates that paper base simulations have been done (although it doesn't say whether these were successful or not).

The next example is from a historical UK defence system, the specific project identification reference information has been protected, but it does seem like a good and useful example anyway.

Table 14.2 Example of 'Full' Hazard Log Entry

Hazard report for XXXXXXXXXX system			
Hazard Number	H27		
Hazard Title	Date used by the system has been corrupted.		
Description	Date received or developed by the system has been corrupted, this may lead to an incorrect decision being made, or incorrect action being taken.		
Accident Title	Fratricide (A3) Equipment loss (A6)	**Accident Severity**	Catastrophic
Project Phase	Build (B)	**Originator**	B Brewer
Use	Operation	**Owner**	J Stewer
Origin	Task Management	**Hazard type**	Functional
Location			
Notes			
Initial Probability	Occasional	**Post Control Probability**	Remote
Current Status	Open		

Status	Change Date	Responsibility	Contemporary probability	Justification (for change)
Open	05 Sept xx	P Gurney	Remote	Introduction of check-sum and alert system ref. TE/1/07.
Open	04 Nov xx	D Whither	Occasional	Initial status entered at safety group meeting SWG03.
Open	12 July xx	T Cobbley	(Unclassified)	Initial Entry of data

You should have picked up that these two logs are considering the same hazard. As you can see, there is significantly more information in the second hazard log entry, giving the opportunity for greater value to be gained from the evidence it contains. There is considerable cross-referencing from the hazard to the accident scenarios, and the inclusion of the person responsible for any changes is kept in the file. This example really does give you a clear picture of the status of this hazard. However, the second hazard log does not contain the explicit mitigation put in place, for that we need to see the external reference TE/1/07 whatever that is. This is perfectly acceptable practice, but there can be significant trailing around to find this information. The cross-referencing is useful, and some might say essential, for safety audit purposes, but there does come a point when you can tie yourself up in knots trying to jump between different documents.

In the US, while the explicit name of a 'hazard log' is not always widely used, the concept of listing and recording hazards is well known and well done. The next example contains hazards on a much wider scale – the whole State of Florida. Many states have a 'Hazard Mitigation Strategy', which lists out the hazards that might have an influence over a wide area and involve whole communities. Its purpose is to [State of Florida 2001];

> ... establish a comprehensive program for the Department of Community Affairs to effectively and efficiently mobilize and coordinate the state's services and resources to make Florida's communities more resistant to the human and economic impact of disasters.

The summary of analysed hazards threatening Florida provides a very useful document to the community, and also matches very well the requirements of a hazard log. The list is as follows:

- Hurricanes and severe storms
- Thunderstorms and tornadoes
- Coastal and riverine flooding
- Freezes
- Wildfires
- Drought
- Agricultural pests and diseases
- Sinkholes and seismic hazards
- Radon
- Technological hazards
- Radiological hazards
- Hazardous materials
- Major transport accident
- Utility or telecommunications failure
- Societal hazards
- Civil disturbances
- Terrorism and sabotage
- Mass Immigration.

The value of looking at existing hazard logs and lists is to give prompts for the hazard analysis of the system, operation or project that you are working on. Someone else may have got a really good and effective mitigation for a hazard you are struggling with. This is also a resource-cheap way of getting a start, although a hazard log should never be copied over, as a significant majority of hazards will just not be appropriate.

Notes

FAA 2000, "FAA System Safety Handbook" Chapter 12, Federal Aviation Administration, US Department of Transportation, Washington, 2000.

ICAO 2006, "Safety Assessment for Operational Trial of BOBCAT" presented at Special Co-ordination Meeting for the Bay of Bengal Air Traffic Flow Management Operational Trial, February 2006.

Leveson, N. G. 2003 "Preliminary Hazard Analysis" White paper, Safeware Engineering Corporation, Seattle, 2003.

MoD 1996, "Safety Management Requirements for Defence Systems Part 2 Guidance", Defence Standard 00:56 Issue 2, Defence Procurement Agency, Glasgow, 1996.

MoD 2004: "Safety Management Requirements for Defence Systems Part 1" Interim Defence Standard 00:56, Issue 3. Ministry of Defence, December 2004.

Railtrack 2000, "Engineering Safety Management Issue 3 Yellow Book 3" Volume 2 Guidance, Railtrack PLC, London, January 2000.

Schäbe, H 2002, "The Safety Philosophy Behind the CENELEC Railway Standards", TÜV Intertraffic GmbH, Köln, Germany 2002.

State of Florida 2001, "The Florida Hazard Mitigation Strategy", State of Florida, Department of Community Affairs, Florida, 2001.

Chapter Fifteen

Human Factors in Safety Cases

Introduction to Human Factors

There are multiple components that make up all but the simplest systems – there may be mechanical components, software, humans, electrical components and the procedures or directions for making the system do its work. There may be implicit laws of physics and engineering, coded directions in a software language or operating procedures from training manuals.

In the construction and operation of a system, we would hope that all the components work together exactly as designed, and that the original design was exactly as required. In our real world, this is not necessarily true. Mechanical components wear out, software can be mis-coded, the procedures may be incorrect or badly taught, the original design may be flawed and even if all this is correct, the human component may still fail!

Perfectly naturally humans get tired (wear out), forget (the simplest) things, have limited abilities of strength and concentration, and will try to make things a little bit easier for themselves. Unfortunately, the design and construction of the other system components can almost 'force' the human to wear out, compromise the human physical ability, dull or swamp the senses or just give too many things to remember at once. These, on their own or in combination, can easily lead into hazardous situations, and hazards can lead to compromises in safety i.e. accidents.

The 'human factor' in system design needs to be understood and optimised according to human ability, such that the system becomes optimised, agile and safe. The human component in a system needs to be given as much attention as the mechanical components, the operating procedures and the management tasks. It has been said that the 'human factor' causes or contributes to anywhere between 60% and 90% of all accidents. Why doesn't it get 60% to 90% of the resources allocated to it?

The Human Caused the Accident

Historical Incident

On May 11, 1996, ValuJet flight 592 crashed into an Everglades swamp shortly after take off from Miami International Airport, Florida. Both pilots, the three flight attendants, and all 105 passengers were killed. Before the accident, the flight crew reported to air traffic control that it was experiencing smoke in the cabin and cockpit. The evidence indicates that five fiberboard boxes containing as many as 144 chemical oxygen generators, most with unexpended oxidizer cores, and three aircraft wheel/tire assemblies had been loaded in the forward cargo compartment shortly before departure. These items were being shipped as company material. Additionally, some passenger baggage and U.S. mail were loaded into the forward cargo compartment, which had no fire/smoke detection system to alert the cockpit crew of a fire within the compartment. On August 19, 1997, the NTSB issued its aircraft accident report entitled "In-Flight Fire and Impact With Terrain; ValuJet Airlines Flight 592." In that report, the NTSB determined that one of the probable causes of the accident resulted from a fire in the aeroplane's Class D cargo compartment that was initiated by the actuation of one or more of the chemical oxygen generators being improperly carried as cargo [FAA 1998].

On the face of it, the cause of this disaster appears to have little to do with human error – there was a fire in the hold, the aircrew could not have done anything to counter it, and the aircraft crashed.

However, investigators learned that several individuals had committed several individual errors over a two month period, each relatively insignificant, but in that combination and in that particular sequence, the disaster was just waiting to happen [Strauch 2002].

The fire was started by oxygen generating canisters that were being carried as company materials (COMAT). They had been installed in another aircraft to provide the emergency oxygen supply in case of cabin air pressure loss, but had gone beyond their use-by date, and so had been removed and were being transported back to the aircraft owners.

In its report on the disaster the National Transport Safety Board (NTSB) uncovered a series of so-called human errors [NTSB 1998]:

1. The oxygen cylinders were not clearly labelled as hazardous materials by the manufacturer or by the maintenance crew, even though there was a general awareness that heat was generated when the cylinders were initiated.
2. Whilst the work card for the task of removing oxygen cylinders did call for the use of a safety cap to be used after removal, this was not done. The work card was signed off as if this task had been completed.
3. During the final inspection of the cargo before it was taken to the aircraft loading ramp, the inspector noticed the lack of safety caps, but was satisfied that "it would be taken care of", he did not check to see that anything had actually been done.
4. The oxygen cylinders were not correctly/securely packaged, labelled or prepared for transport, enabling them to be free to move about.
5. The potentially hazardous content of the shipment box was not communicated about between maintenance staff, storage staff, ramp-loaders and flight crew.

Other contributory factors were also highlighted [ibid.]:

1. There had been a push for smoke detection and deluge equipment to be installed in aircraft holds. The FAA terminated the rule-making action to require

such systems citing that these systems were not cost beneficial, and that they would not provide a significant degree of protection to the occupants.

2. Safety equipment was provided to the aircrew in the form of oxygen masks and smoke goggles. Emergency procedures stated that he crew should don these as soon as smoke is reported. There was no evidence that this had been done – the voice recordings were all clear (i.e. un-muffled), and there was evidence that the crew were proceeding with smoke clearance actions. It was noted that the plastic packaging for this emergency equipment was of a strong nature that usually required both hands or a sharp implement to actually open.

Many points of critical interest come from these findings. Unfortunately, it is likely that if just one of the 'human factors' errors had been identified the cargo would not have been loaded. The regulator (FAA) appears not to have understood all the purposes of the proposed smoke detection and deluge systems; earlier warning may have given enough time for this flight to return and evacuate the aircraft, even with the hold fire progressing. Finally, even with all the concentration on the human factor, consideration of the small detail of the emergency equipment packaging might have given extra time. Tragic.

Historical Incident

A demolition firm was ordered to pay £43,000 in fines and costs after a court heard how a four-year-old child was seriously injured when a 20-tonne loading shovel vehicle [digger] rolled down a hill and tipped over. The driver was part of a team carrying out landscaping work at a housing estate demolition site. On the day of the accident, the driver was using the shovel when a colleague asked for assistance with his task. The driver left the vehicle with the front scoop elevated and loaded when he parked it. This caused it to roll away and tip over when it hit the road edge. The loading shovel injured two children aged four and five, who were playing near the site, as well as the driver who tried to stop it from toppling over. One child suffered serious crush injuries with two broken legs. The court heard that the driver had not received any training on how to use the loading shovel correctly. In addition the demolition site had not been adequately cordoned off to prevent children from playing close to the works [BSC 2006].

This incident highlights further areas where human-based contributions at different levels can conspire to cause a critical accident. Three points of interest come from this report. Firstly, the equipment safety awareness of the driver – leaving a loaded shovel in the elevated position (it was easier than unloading). Secondly, the lack of training provided to the driver – probably the responsibility of someone far removed from the site. Lastly, the failure of the cordoning-off procedures – again probably not the responsibility of anyone directly involved in the accident.

James Reason quotes a social scientist when discussing the way that multiple defences intended to give depth to accident prevention are sometimes easily defeated in unimagined ways [Weick in Reason 1997];

We know that single causes are rare, but we don't know how small events become chained together so that they result in a disastrous outcome ... to anticipate and forestall disasters is to understand regularities in the way small events can combine to have disproportionately large effects.

Reason has developed the 'Swiss cheese model' to simple demonstrate the way that there are a series of holes in all the precautions that are taken. On the rare occasions that these holes line up, or occur together, the precautions are breached and the accident happens [Reason 1997]. The human is certainly one, and sometimes more, of the holes in the cheese model, with mechanical systems, software etc. being the others.

However, it is also true that the human does make up a lot of the cheese as well.

The Human Prevented the Accident

I will start this section with a familiar historical incident, but provide different commentary on it to demonstrate the point behind the sub-title.

Historical Incident

A demolition firm was ordered to pay £43,000 in fines and costs after a court heard how a four-year-old child was seriously injured when a 20-tonne loading shovel vehicle [digger] rolled down a hill and tipped over. The driver was part of a team carrying out landscaping work at a housing estate demolition site. On the day of the accident, the driver was using the shovel when a colleague asked for assistance with his task. The driver left the vehicle with the front scoop elevated and loaded when he parked it. This caused it to roll away and tip over when it hit the road edge. The loading shovel injured two children aged four and five, who were playing near the site, as well as the driver who tried to stop it from toppling over. One child suffered serious crush injuries with two broken legs. The court heard that the driver had not received any training on how to use the loading shovel correctly. In addition the demolition site had not been adequately cordoned off to prevent children from playing close to the works [BSC 2006].

In looking at the title of this section and with regard to the previous section, you might be wondering why this incident has been brought up again. Well, what stopped this critical accident from turning into a fatality? The driver was injured in the accident trying to stop the vehicle from tipping over, but hang-on, he was already out of the vehicle to give assistance to a colleague, how was he hurt? This is not expressly recorded, but it is my proposal that he noticed the vehicle running away and ran after it to try and stop it. He probably would have been shouting a warning to the children, who looked and perhaps tried to get out of the way. The actions of the driver probably prevented the death of both children – the human had prevented a worse severity accident.

Further records of this type are hard to come by – near misses are frequently not recorded in the same way, they don't come to court. But there are frequent records of people being saved from drowning, being saved from fires, indeed this is the whole purpose of a great deal of government funded emergency services. Also, I am pretty certain that you will have heard, or even said yourself "Hey! Watch out".

Accident severity studies have shown the occurrence ratio between serious injury accidents, minor injuries and incidents with no reported injury or damage. One such study [Bird & German in Engineering Council 1993] cites that the ratio is 1-to-10-to-600. This concept has become known as the ice-berg principle, where each increasing severity class is considered an order less likely to occur. This is somewhat supported by the incident data discovered during the cost-benefit studies referenced in Chapter 12, showing typically single figure serious incidents and tens or hundreds of other incidents with very little impact.

Consider the front-shovel incident again, in this instance a human action appears to have prevented a death from occurring, keeping the incident as a serious injury rather than a fatality. This phenomenon is likely to repeat throughout the accident severity ratio. OK, many accidents are only ever going to be minor because they involve low energy values (see Chapter 13 on hazardous energy analysis), e.g. a stapler and some paper. However, a significant proportion of incidents do have their severity reduced, or are avoided altogether specifically because there was a human operator, technician or by-stander at the scene who saw and understood what was happening and deliberately intervened in the accident sequence.

So whilst human error is regularly blamed for x% of accidents, I believe that human ability is responsible for preventing 100 times the incidents that there would be if the human component was not involved in the system. The human as the cause is an important issue to be minimised, but the human as the preventer is possibly more important and should be maximised. The human should not be made quite so distant from the control system as is often believed.

The impact of removing the 'unreliable' human from the system has become known as 'The ironies of automation' [Bainbridge in Reason 1997]. There are several points to highlight based around the idea that by taking away the easy parts of the operator tasks, automation makes the difficult parts much more difficult.

- Many system designers see the human as unreliable and inefficient, yet they still leave the human to cope with the tasks that they couldn't work out how to automate – most especially the restoring of a system to a safe state after some failure.

- In highly automated systems, the task of the human operator is to monitor the system to check that the automated parts are working. But it is well known that even the most motivated individual has trouble maintaining concentration and vigilance over long periods of time. Thus, humans are ill-suited to watch out for rare abnormal conditions.

- Skills need to be practised regularly in order to maintain them, yet an automated system that fails very rarely denies the human the experience to practice skills that would be essential if/when the emergency occurs. So there is a much greater need for regular, specialised (expensive) training.

Human Systems Integration

The human is a key component for the safe operation of many systems, their impact on the system has to be recognised and understood for any safety case to be successful. The complexity of technological systems has increased rapidly over the last century, and is increasing even as you read this book. On the other hand, the human has remained evolutionarily fixed for the last million years or so. With technology starting to expose the limitations of even the best human operators, the 'human factor' has to be integrated into the system. It can not be considered as an external component that can work it out when given something new or innovative.

Reducing hazardous incidents is just one of the benefits of good human systems integration. Products with good integration have lower product recalls, reduced liability, lower development costs, better time-to-market, lower training needs and reduced support costs [MoD 2005]. Good human systems integration also means that a wider proportion of the population will be able to use the system/equipment that is being developed. If more people can use it, it will be better for the market place. If more workers can use the system safely, there will be a wider pool of potential recruits to tap into i.e. not expensive specialists.

In order to appreciate human systems integration, it is advisable for all projects to carry out some early stage human factors analysis. This can identify issues that are going to be important later in the project, and can allow trade-offs between design and safety areas with a good explicit knowledge of the human issues. In general an early human factors analysis can be carried out using a three-stage process as follows [ibid.]

> Step 1: Identification and collation of relevant information in readiness for the integration analysis. This might include contemporary standards, assumptions, expressions of system requirements, comparisons with earlier or similar systems in existence and any performance targets or limitations that have been established.
>
> Step 2: Assessment and prioritisation of the issues in readiness for reporting. The issues may be rated according to the likelihood of occurrence and their potential impact to the project [Does this sound familiar?]. This may be done in a workshop or working group process, with all resulting issues, safety and others being recorded in a human factors issues register or log.
>
> Step 3: The risks from the assessment are formally recorded and strategies for reduction of the risks are proposed. The output document should give conclusions and recommendations to the project leaders. This analysis report is likely to be important evidence for higher level human factors documents and ultimately the safety case itself.

The US Army was the first large-scale organisation to fully implement and demonstrate the benefits of an human systems integration approach [Booher

2003].   In 1986 the US Army created a manpower and personnel integration management and technical programme called MANPRINT, it was designed to improve military systems and performance.   The most important aspect of the programme was the integration of human factors into the main system-engineering domain.   The programme has integrated multiple fields of military operations covering manpower, personnel, training, human factors engineering, system safety and health hazards.   After numerous fratricide incidents the additional field of 'soldier survivability' was added [ibid.].   The MANPRINT approach has now carried over in to many other industries and countries – e.g. the UK MoD, the US FAA and The Netherlands Applied Scientific Research Organisation to name just a few.

Safety Documents from the Human Factors Domain

I have already introduced the first document that will be useful for use in developing a safety case, the early human factors analysis report.   But this leads on to further, more refined documents as the project or system develops over time.

Human Systems Integration Plan

The phrase 'Human System Integration' may be taken as the rest-of-the-world equivalent to the UK phrase 'Human Factors Integration'.   It's another nomenclature problem that exists in the English speaking world. The addition of the word 'Plan' does change the meaning, and it becomes quite specific.   For example, the US Department of the Navy recommended format and content of the Human Systems Integration Plan (HSIP) is as follows [DoN 1996];

> Executive Summary
> Introduction
>   Objectives and scope
> System Description
>   General description of the system
>   Major system components
>   Performance characteristics
>   Performance goals and thresholds
> Issue and Constraints
>   Manpower issues
>   Manpower availability
>   Human capability and training issues
>   Human performance issues
>   Systems safety, health and environmental issues
> HSI Program
>   HSI Objectives
>   HSI Strategy
>   HSI Analyses
>   HSI Analyses results
>   HSI Test and evaluation
>   HSI Relationships

HSI Activities
Annexes covering points of contact, references, data sources, issues and history
log.

This is not a consistent description fitting all HSIPs, but it is considered indicative
enough to provide a good baseline. There is a considerable synergy between
HSIPs and safety plans – the system description will be consistent, the issues and
constraints are of high interest, and even the annex on issues and history log will
help to provide evidence to the hazard log.

Human Factors Integration Plan

In the UK, one name for the planning document is 'The Human Factors
Integration Plan' (HFIP), and it aims to provide a co-ordinating document for all
integration activities within the project. In particular, it should include the
following type of information [MoD 2005]:

HFI issues
HFI constraints
HFI studies, actions and mitigation strategies
Allocation of actions between MoD and industry
Dependencies between organisations and project activities
Strategy for integrating HFI data collection
Key milestones, deliverables and time scales
Extent of need for end-user participation
Method for including HFI in system level trade-offs
Method for monitoring and controlling progress of the plan.

There is certainly some carry-over of information from this work into the safety
case domain – the issues and constraints are of most interest, and it is likely that
the information exchange will be both ways – as it should be. The advice does
not appear to be as well thought out and as mature as the US example – perhaps
there are quick lessons for the UK available, perhaps the two will converge over
time.
    The final piece of advice here is to emphasise that the document set required
for safety analysis and safety case production relies on the valuable work and
evidence provided by the whole of the project or operational staff and not simply
the work of the designated safety team. We are not at home to Mr Stovepipe.

Human Factors Analysis

There is a significantly large suite of tools and techniques for human analysis,
most of which can provide important data and evidence to a safety case. Some we
have already seen in Chapter 13, there are plenty more and the chances are that
the safety engineer will have to be aware of them and what they can deliver to the
safety case.

NASA gives some excellent summary advice on the tools and methods used by human factors practitioners, you may be lucky enough to be involved in some of them as a safety specialist advisor:

Cognitive walkthrough
Cognitive walkthrough requires an HCI practitioner to ask a set of usability questions regarding a proposed interface. Cognitive walkthrough requires a well specified prototype and task as it is designed to analyze each step in a task to determine whether or not the user is likely to succeed. For example, does the user have the right goal and will the user notice the correct action is available?

Contextual inquiry
Contextual inquiry is an adaptation of ethnographic methods for HCI. The inputs to the methods are similar to ethnography. It places the HCI researcher in the role of apprentice in order to observe and participate in the user's real work. The outputs of the method are engineering style, box and arrow diagrams that capture and communicate the various aspects of the work (e.g. information flow, physical environment, etc.).

Ethnography
Ethnography is a staple of cultural anthropologists that places the researcher squarely in the research context as a participant-observer. The broad focus uncovers issues not directly tied to user goals or tasks but that could put the project at risk if not understood. For example, failing to understand data security is important to a group of science users even if the set of tasks appears public.

Heuristic evaluation
Heuristic evaluation requires an HCI practitioner to compare a given interface to a set of heuristics or best practices that capture the key point elements of successful interfaces. One heuristic is, 'does the system match the real world?' For example, does the interface terminology match the user's terminology or does it use system centric terms? Heuristic evaluation does not require a specified task and is often considered a discount usability method because it is quick to perform and identifies a large number of problems.

Human Performance Modeling: Model Human Processor
The human performance modeling methods generate a priori predictions about time on task for skilled human operators doing routine tasks. This family of methods relies on the composition of tasks from low operators (e.g. a cognitive cycle to send the command to a hand or eye is milliseconds according to the literature).

Personas
Personas are a way of synthesizing a concrete individual instead of an 'average user' that is difficult to conceptualize and design for. Often a persona is developed for each type of user, and one is highlighted as the hardest to serve. For example, between the flight attendant, business traveler and vacationer the hardest to serve is the vacationer because they will never become an expert user of the system. Personas require their designers to get very specific about who they are from their computer experience to the type of car they drive.

Process Analysis

Process analysis focuses on the larger work process a software application fits into. Ideally, process analysis is done as part of user research. If this is not possible, (e.g. the process has not been established or there are no users) process analysis can be done as a system is being prototyped or after it is in production such that the next version is adapted to fit the process. Based on an understanding of the existing software tools, procedures, and user roles, it can also be effective to propose process change which is often less costly and can be applied more quickly.

Prototypes

Prototypes are mockups of a proposed interface. The purpose of a prototype is to simulate user interaction, often in the context of a user test, with a system that does not yet exist. Based on the kind of data required and time allotted, different types of prototype are appropriate. For an early stage, informal discussion low fidelity paper or an HTML storyboard are often effective. For a detailed test of a highly interactive system, a high fidelity prototype is necessary, built in a tool like 'Director' or 'Flash'. Effective prototyping requires an iterative design and evaluation cycle.

Scenarios

Scenarios are created based on user research. Designers will have identified a set of common tasks. Scenarios are specific examples of these tasks. Often designers will test their early stage interface concepts against a set of scenarios to make sure those particular tasks can be completed with the proposed interface.

Storyboards

Storyboards are a visual representation where a step or action in a task is represented by a set of consecutive panels much like a comic book. This method is a low fidelity way to flesh out the interaction design required to support an existing scenario and effectively communicate it to users or stake holders.

Surveys

Surveys used in HCI are conducted according to standard research design methods. Their purpose is often to gather data about user populations or measure user satisfaction in a quantitative manner.

Task Analysis

Task analysis is a method used to understand the user's current task at a fine grain size. Task analysis often involves a hierarchical decomposition that can be used independently to redesign the task or as input to another method like human performance modeling.

Think Aloud Usability Tests

Think aloud usability tests involve users of the system working to complete a well specified task with a prototype or implemented system. The HCI practitioner then analyzed the data collected during the user test, often as video or notes and identified critical incidents based on criteria identified beforehand. These criteria are based on project needs and can vary such that on one project spending 5 or more minutes to recover from an error is a critical incident. On another project a critical incident may be spending 30 or more seconds.

Use Cases
Use cases are a more rigorous, engineering style version of scenarios. They focus on information exchange between a system and a user by specifying who the user is, exactly what information they have, and the system's response at every stage.

There are now may excellent text books discussing the tools and techniques of the human factors/human systems domain e.g. Reason's Managing the Risks of Organisational Accidents, Kirwan's Guide to Practical Human Reliability Assessment and the UK Health and Safety Executive guidance (HSG48) on reducing error and influencing behaviour. In the UK a new suite of defence standards, 00:25, has been released giving voluminous information about human factors for designers of systems. If you want more information on this subject please refer to these texts.

Summary

The human is an integral component of the system, their performance and capability directly affect the performance and capability of the whole system. It is essential to acknowledge, analyse and record the reliability of the human component in the system safety case.

The human will contribute to the risk profile of the system, both as an event initiator and, if you are clever, as an event prohibitor.

Notes

Booher 2003, "Handbook of Human System Integration", John Wiley & Sons Inc, New Jersey, 2003.

DoN 1996, "Defense Acquisition Deskbook", US Department for the Navy, 1996.

Engineering Council 1993, "Guidelines on Risk Issues", The Engineering Council, London, 1993.

FAA 1998, "RIN 2120-AG35 Prohibition on the Transportation of Devices Designed as Chemical Oxygen Generators as Cargo in Aircraft" Notice of proposed rulemaking 98-12, Federal Aviation Administration, Department of Transportation, 1998.

MoD 2005, "The MoD HFI Process Handbook" MoD human factors integration defence technology centre, MoD Defence Procurement Agency, Bristol, 2005.

Reason 1997, "Managing the Risks of Organisational Accidents", Ashgate Publishing Limited, Aldershot, 1997.

Strauch 2002, "Investigating Human Error: Incidents, Accidents and Complex Systems", Ashgate Publishing Limited, Aldershot, 2002.

Chapter Sixteen

Software Factors in Safety Cases

Introduction to Software Factors

Software programs are used in many applications from weapon control systems, communication equipment, and flight control systems to medical support machines, banking and car engine management systems. Computer systems are used to perform a variety of essential non-safety functions, safety related functions and even safety critical functions ('safety critical' is an accepted industry standard phrase to be used where the failure of the computer program can lead *directly* to a fatality).

Any safety assessment work must consider any existing software, as a potentially equal source of risk when compared to the equipment and people involved. It is often the case that a specific software safety case is called for where a particular system contains a software intensive product.

The Software Caused the Accident

The title to this section is not strictly true, the failure of software is not directly hazardous, however hazards arise from inappropriate computer-based control of a system and/or the presentation of hazardously misleading decision support information. The very idea of software *failure* is almost a misnomer – software does what it has been told to do, it doesn't wear out, fracture or breakdown in a sort of normal statistical distribution of failure likelihood. As far as the software is concerned it does exactly what has been asked of it, it hasn't *failed*. As far as the system is concerned, the software can fail to provide the required capability, causing the system to fail. This is a failure to satisfy the system requirements, OK so software has been used as the tool, but the real failure has been in the writing (or coding) of the software, not in the software performance.

Failures of computer systems arise from systematic causes – flaws in specification, flaws in design implementation or unanticipated use influences. By their nature, systematic flaws are not known in advance, every historic or future design action could be responsible for them, so it is very difficult to find and remove them.

Historical Incidents [Garfinkel 2005]

In October 2005, automaker Toyota announced a recall of 160,000 of its Prius hybrid vehicles following reports of vehicle warning lights illuminating for no reason, and cars' gasoline engines stalling unexpectedly. But unlike the large-scale auto recalls of years past, the root of the Prius issue wasn't a hardware problem, it was a programming error in the smart car's embedded code. The Prius had a software bug.

1985-1987: Therac-25 medical accelerator. A radiation therapy device malfunctions and delivers lethal radiation doses at several medical facilities. Based upon a previous design, the Therac-25 was an "improved" therapy system that could deliver two different kinds of radiation: either a low-power electron beam (beta particles) or X-rays. The Therac-25's X-rays were generated by smashing high-power electrons into a metal target positioned between the electron gun and the patient. A second "improvement" was the replacement of the older Therac-20's electromechanical safety interlocks with software control, a decision made because software was perceived to be more reliable. What engineers didn't appreciate was that both the 20 and the 25 were built upon an operating system that had been kludged together by a programmer with no formal training. Because of a subtle bug called a "race condition", a quick-fingered typist could accidentally configure the Therac-25 so the electron beam would fire in high-power mode but with the metal X-ray target out of position. At least five patients die; others are seriously injured.

January 1990: AT&T Network Outage. A bug in a new release of the software that controls AT&T's #4ESS long distance switches causes these mammoth computers to crash when they receive a specific message from one of their neighboring machines. Its a message that the neighbors send out when they recover from a crash. One day a switch in New York crashes and reboots, causing its neighboring switches to crash, then their neighbors' neighbors, and so on. Soon, 114 switches are crashing and rebooting every six seconds, leaving an estimated 60 thousand people without long distance service for nine hours. The fix: engineers load the previous software release.

1993: Intel Pentium floating point divide. A silicon error causes Intel's highly promoted Pentium chip to make mistakes when dividing floating-point numbers that occur within a specific range. For example, dividing 4195835.0/3145727.0 yields 1.33374 instead of 1.33382, an error of 0.006 percent. Although the bug affects few users, it becomes a public relations nightmare. With an estimated 3 million to 5 million defective chips in circulation, at first Intel only offers to replace Pentium chips for consumers who can prove that they need high accuracy; eventually the company relents and agrees to replace the chips for anyone who complains. The bug ultimately costs Intel $475 million.

June 4, 1996: Ariane 5 Flight 501. Working code for the Ariane 4 rocket is reused in the Ariane 5, but the Ariane 5's faster engines trigger a bug in an arithmetic routine inside the rocket's flight computer. The error is in the code that converts a 64-bit floating-point number to a 16-bit signed integer. The faster engines cause the 64-bit numbers to be larger in the Ariane 5 than in the Ariane 4, triggering an overflow condition that results in the flight computer crashing. First, the Flight 501's backup computer crashes, followed 0.05 seconds later by a crash of the primary computer. As a result of these crashed computers, the rocket's primary processor overpowers the rocket's engines and causes the rocket to disintegrate 40 seconds after launch.

November 2000: National Cancer Institute, Panama City. In a series of accidents, therapy planning software created by Multidata Systems International, a U.S. firm, miscalculates the proper dosage of radiation for patients undergoing radiation therapy. Multidata's software allows a radiation therapist to draw on a computer screen the placement of metal shields called "blocks" designed to protect healthy tissue from the radiation. But the software will only allow technicians to use four shielding blocks, and the Panamanian doctors wish to use five. The doctors discover that they can trick the software by drawing all five blocks as a single large block with a hole in the middle. What the doctors don't realize is that the Multidata software gives different answers in this configuration depending on how the hole is drawn: draw it in *one route [around the node points] and the correct dose is calculated, draw in another direction and the software recommends twice the necessary exposure.*

At least eight patients die, while another 20 receive overdoses likely to cause significant health problems. The physicians, who were legally required to double-check the computer's calculations by hand, are indicted for murder.

In one final example, the crash of a DC-10 into Mt. Erebus in Antarctica, the software is not blamed, but the data in the computer system for the navigation is. The aircraft in question was a tourist flight giving sight seeing tours around Mt. Erebus, to assist with finding the 'sights' in Antarctica, the auto pilot has the exact co-ordinates of the mountain plotted in. Due to weather conditions, there appeared to be a loss of spatial awareness and the plane hit the mountain. However, Justice Mahon said that the "... the single dominant and effective cause of the disaster was the mistake made by those airline officials who programmed the aircraft to fly directly at Mt. Erebus and omitted to tell the aircrew." [Mahon 1981].

From the above historical incident examples and discussion, the software code itself is not always the problem. However, as far as any safety case is concerned, the exposure to risk from the computer system as a whole is justifiably highlighted throughout. The critical point to take from this is that there is a widespread prejudice that software is the nasty component. As such, many incidents are blamed on software bugs or failures, and reports and meetings have a tendency to label all computer system failings as 'software failures'. This simply is not true. I have heard it expressed that a so-called software error is in-fact a human error further up the supply chain i.e. during coding, design or requirements capture.

At this point, it is worth going back over the incidents noted above to review where the failure actually was – the requirement specification, the design of the software code or through unanticipated use conditions.

Commercial-off-the-shelf (COTS) Systems

Commercial off the shelf systems are those that are bought or procured for the functionality that they already have. They are systems that are not designed with a specific single project in mind – they are not bespoke. The Microsoft operating system is a commercial off the shelf system, it has multiple functions, some of which any particular project might need. Not all the functions are needed, but they come with the mass-produced package, so all functions are procured. This is perceived as being a much cheaper way of getting the functionality that a system is likely to require.

COTS products have the following characteristics [O'Halloran 1999]:

- They are readily available from commercial sources
- They have general rather than specific applications
- They are not designed for a specific environment
- Access to detailed code data is usually denied
- Configuration control is usually suspect

- They are usually not modifiable
- Some elements have a technological support lifetime of less than 2 years.

These factors usually mean that there is a high refresh rate of computer-based systems in many industrial and military domains. In the military domain specifically, where the expected life-time of any system can be 20 years or more, technological advances can reduce or remove the military advantage that was anticipated at the initial procurement time. This in turn means that the safety analysis and safety case construction, where COTS products are continually being replaced with the next generation of COTS products, can be a similarly continuous requirement.

There are acknowledged to be three ways of dealing with COTS systems (including software code alone) in safety analysis and safety cases [ibid.].

1. Ignore it: if COTS systems are ignored, they are being treated as trusted components, but without supporting documentation and evidence. The safety claim and argument built on these non-existent foundations will inevitably fall.

2. Analyse and test it: this is certainly possible to do once the computer system has been purchased, although if it turns out to be full of errors and unusable, it can be very expensive and there is still no product to use. There are several automatic system testing methods available. These are usually owned by specific commercial organisations, or are only available to government departments, but they do make significant findings.

3. Isolate it: treat the computer system as untrustworthy and ensure that it does not come into contact with the safety critical functionality of the whole system. However, this does mean that some additional bespoke component has to be developed for the safety critical parts of the system. Also, the interfaces between the critical and non-critical system components have to be managed and defined precisely, so that they can be formally proven to be correct.

Software of Uncertain Pedigree (SOUP)

Software is the major component of computer-based systems, sometimes bespoke software is specifically developed for a particular project. Very often though, a significant amount of the software used will be re-use of existing software e.g. operating systems, compilers, user interfaces, system libraries etc. The software may have been designed specifically for safety related tasks, more likely it will have been developed for commercial reasons. However, the approach does have advantages over bespoke software development. Development and coding time is certainly reduced (even to as low as nil), there may be extensive prior use in other similar environments (a rich source of evidence), and of course it is likely to be cheaper [HSE 2001]. An analysis of the trouble, time and expense to create bespoke software for a safety related risk situation, may lead to that course of action being beyond grossly disproportionate to the risk. The same analysis for

SOUP might demonstrate different findings, better justifying its use over bespoke code. This selection can only be done rigorously of course, if the analysis is done, and naturally this should be recorded in the safety case documentation.

How to Treat the Risks of Software

The requirements to manage the risk of the system remain for software and computer-based components, but as the risks are caused through systematic (not statistically random) fault mechanisms, the actual analyses and evidence provision tasks are often specific to this type of system. A good practice recommendation from the UK Health and Safety Executive does indicate that the software failure process arises from the random uncovering of the systematic faults during execution of successive inputs. This uncertainty can only be captured by probabilistic representations of the failure process. Hence, the use of probability-based measures to express our confidence in the reliability of the computer system is therefore inevitable [HSE 1998].

Many standards on the performance of complex electronic systems have introduced the concept of 'safety integrity levels' (or similar equivalents e.g. design assurance levels or safety assurance levels) e.g. IEC 61508, RTCA DO178B, Def Aust 5679. The safety integrity levels (SILs) provide an indication of the required level of risk protection on a system, they indicate the degree to which a complex electronic system component should be free from flaws. These standards explicitly define good practice for each of the levels and therefore implicitly link engineering methods and tools directly to the level of risk. In having this link in place, a strategy for achieving the integrity level is prescribed [MoD 2004].

In some domains this can work very well, it has the advantage of providing an authoritative definition of good practice and makes it easy to conform to the standards in question – 'do this and you pass'. However, technology moves faster than standards, and there can be problems conforming when innovative or novel systems are employed, e.g. integrated modular avionics for aircraft (or other vehicle) control systems. Additional problems can arise if the SIL-based scheme is applied outside its specific domain, or if systems cross-over between domains e.g. where civilian vehicles are used for military purposes.

The UK has evolved some of its standards in this area to be more goal based. This is where the result or evidence requirements are specified, but the methods and tools are not. The method and tool selection is left to the discretion of the project engineers and managers, based on the anticipated level of risk, the required level of confidence and the abilities, preference and availability of support and staff. This methodology does require more effort to understand and plan the software risk management, but it does give more flexibility to the project, which may (only 'may' and not 'will'), reduce costs and improve time to market.

Safety Integrity Levels

The main aim of the safety integrity levels concept is to ensure that complex programmable systems that carry out safety functions provide the required level of safety performance. To achieve this, it is first necessary to determine what is this required level of safety integrity. This has to be done through an analysis of the hazards and risks associated with the system and the developed acceptable failure rates, which can be tolerated from the system. In the IEC 61508 standard, the allocation process is described in two tables 16.1 and 16.2, based on the units of measurement used to specify the target failure measure (See chapter 8 for a discussion on units) [IEC 1998-2000].

Table 16.1 Safety Integrity Levels for On-demand Function

Safety integrity level	Average probability of failure to perform design function on demand
4	10^{-5} to 10^{-4}
3	10^{-4} to 10^{-3}
2	10^{-3} to 10^{-2}
1	10^{-2} to 10^{-1}

Table 16.2 Safety Integrity Levels for Failures per Hour

Safety integrity level	Probability of dangerous failure per hour
4	10^{-9} to 10^{-8}
3	10^{-8} to 10^{-7}
2	10^{-7} to 10^{-6}
1	10^{-6} to 10^{-5}

Consistent across the computer safety standards that have adopted the integrity level concept, is a concept where a specification is made on the system development, language, control and testing processes for each integrity level. A typical example from the Motor Industry Software Reliability Association (MISRA) is given in table 16.3 [MISRA 1995].

Claims of compliance with a particular safety integrity level can then be made providing the evidence of conformance is recorded and made available, i.e. you actually have to have done the work as specified in the compliance table.

Required Level of Confidence

In setting safety requirements for software and computer-based systems, and then when assessing if they have been met, it is vital to acknowledge that there can be no absolute guarantee that a system meets its safety requirement. There can only be greater or lesser confidence that this is the case [MoD 2004]. It is therefore important to consider how much confidence is actually required. The integrity level indicates the level of confidence there should be in the provided evidence for the safety case. The higher the integrity requirement, the higher the confidence should be. Guidance is provided in the UK through the MoD standards for safety management of defence equipment [ibid.], this is summarised in Table 16.4.

Table 16.3 Example of Compliance Actions for Differing SILs

Developme nt process	1	2	3	4
Specification	Any structured method.	Structured method supported with tool.	Formal specification.	Formal specification plus auto-code generation.
Language & Compiler	Standardised structured language.	Restricted subset of a standardised structured language.	As for level 2.	Independently certified compiler with proven formal syntax and semantics.
Configuration Controlled Products	All software products & source code.	Relationships between all software products & all tools.	As for level 2.	As for level 3.
Configuration Control Process	Unique identification, product matches documentation. Access control.	Control and audit changes. Confirmation process.	Automated change and build control. Automated confirmation process.	As for level 3.
Testing	Show fitness for purpose, test all safety requirements.	Black box testing.	White box testing – defined coverage. Stress testing against livelock and deadlock. Static analysis.	100% white box module testing. 100% requirements testing. 100% integration testing Semantic static analysis.
Verification & Validation	Show tests are suitable, exercise safety features, traceable correction.	Structured program review. Show no new faults after correction.	Automatic static analysis, proof of safety properties. Justify test coverage, show test are suitable.	All tools to be formally validated. Proof of code against specification. Show object code reflects source code.
Access for assessment	Requirements and acceptance criteria. QA and product plans. System test results.	Design documents, software test results, training structure.	Techniques, processes, tools. Witness testing, adequate training. The Code.	Full access to all stages and all processes.

Table 16.4 Levels of Confidence and Suggested Evidence

Level of confidence	Suggested evidence
HIGH The highest level of confidence possible given the [contemporary] state of the art.	Diverse forms of evidence, each providing high confidence should be combined. Argument should be based on integrating evidence from testing, analytical arguments and examinations together with qualitative arguments of good practice and process metrics. Evidence should be rigorous and comprehensive (i.e. 100% coverage in testing regimes). The functionality, performance and behaviour of the system should be understood in its entirety in its operating context. For COTS components, the co-operation of the supplier is likely to be essential, as design and manufacturing specifications and data will be required. All the evidence should be made available for rigorous independent audit and scrutiny.
MEDIUM The effort expended on providing confidence should be proportionate to the risk.	The evidence may rely on a single primary argument with high confidence plus some supporting evidence, or on the combination of several arguments, which can each have a lower level of confidence. Evidence should include analytical, qualitative and quantitative data, although coverage need not be 100%, provided the effort is directed towards areas of greater risk. For COTS components, supplier co-operation is normally necessary, but if comprehensive examination is practicable this may not be essential. Evidence should be independently reviewed, but a sampling approach may be used.
LOW	The evidence need only come from testing, field data and demonstration of compliance with standards. For COTS components, design and manufacturing data is not necessary, although some evidence of good engineering practice needs to be demonstrated. A sampling approach may be used for auditing.

Software Testing Methods

There is a significantly large suite of tools and techniques for testing software, most of which can provide important data and evidence to a system safety case. It can be difficult to find recommended best practice for every situation, as new situations come along surprisingly quickly. Even in just a few more years, some of this section (and the one on human factors analysis techniques) will need to be updated as new techniques are developed. From a safety perspective, there should be understanding and agreement as to the tests, processes, timing, tools and rigour that are going to applied to a software or computer based system.

It is beyond this text to give pages of details on a multitude of techniques, however a summary of some of the techniques is given here cribbed from the US FAA system safety handbook [FAA 2000]. This cites that the following techniques may be used, particularly during the software design phase:

* Design logic analysis
* Design data analysis
* Design interface analysis
* Design constraint analysis
* Software fault-tree analysis
* Petri-nets
* Dynamic flowgraph analysis
* Use of safe subsets of program language
* Formal methods
* Requirement state machines.

IBM has also carried out some research on software development best practice citing that:

> Every time we [IBM] conclude a study or task force on the subject of software development process, there is one recommendation that comes loud and clear, "We need to adopt best practices in the industry." While it appears as an obvious conclusion, the most glaring lack of it's [best practice guidance] presence continues to astound study teams. [Chillarege 1999]

The IBM list is broken down into three sections *basics* – the well known, broad-based starting position for testing; *foundational* – not so well known but valuable supporting techniques; *incremental* – for specific advantages under special conditions.

The Basic Practices
* Functional specifications
* Reviews and inspections
* Formal entry and exit criteria
* Functional test – variations
* Multi-platform testing
* Automated test execution

- Beta programs
- Nightly builds.

Foundational Practices
- User scenarios
- Usability testing
- In process ODC feedback loops
- Multi-release butterfly profiles
- Requirements for test planning
- Automated test generation.

Incremental Practices
- Automated environment generator
- State task diagrams
- Memory resource failure simulation
- Statistical testing
- Formal methods
- Check-in tests for code
- Benchmark trends
- Bug bounties.

For more details on these techniques, the specific cases where they are used and the evidence that they produce, please see the references.

Software testing is also broken down in to two broad types – white-box testing and black-box testing. White-box testing is where the design of the test uses the architecture of the system as the driver of the test case. It will check all independent paths at least once, exercise all loops at their boundaries, test the internal data structures and trial all logic-based decisions. To carry out white-box testing, the internal construct of the system needs to be known explicitly (generally not available with COTS products). Black-box testing relates to tests that are done at the interfaces between the components, testing the inputs and outputs with little interest in the actual way the computer system is internally structured.

Black-box testing techniques are not to be considered as an alternative to white-box techniques – they should be used as complementary approaches, as each will uncover different types of flaws. This is the real key to software testing, a single approach is unlikely to find all (and sometimes any) of the flaws in a computer and software based system. A combination of techniques is essential in order to demonstrate that a sufficiently rigorous approach has been carried out and that the correct level of confidence in the performance has been achieved.

Note however, that it is very difficult to ensure that there are not common mode failures in software. This means that where software is always required to deliver a function or service, all the software connected with that hardware device needs to be shown to have the same (high) level of integrity.

Safety Documents from the Software Domain

A significant document suite is recommended for software, most of it is likely to be specific to the software development processes used. There are several of these documents that will be of great interest to the safety engineering team, indeed, software and safety should be significantly interwoven when it comes to producing some of the software safety evidence, allowing greater exchange of concerns, analysis, ideas and mitigation techniques.

As ever, there is not a consistent set of titles of what these documents may be called in each country, industry or company, and I am afraid I don't know them all. The description and purpose of the documents are more important than the title, but I know that project staff talk in acronyms and abbreviations, so I hope the labels I have used are not misleading.

Software Safety Case

The software safety case presents a readable justification that the software is safe to use for its intended purpose [MoD 1997]. The contents should include:

- Description of the system configuration and software versions
- Summary of the design approach to safety
- Description of the software safety properties
- Description of the software architecture
- Description of the arguments used to demonstrate requirement satisfaction
- Description of the development process, methods and tools
- Justification for the inclusion of any COTS components
- Description of any outstanding issues
- Summary of any changes to the software safety case
- Analysis of compliance with applicable software standards
- Description of any field experience of any part of the system
- Summary of the process evidence used in the safety argument
- Summary of the direct testing evidence used in the safety argument
- Summary of any counter-evidence discovered and the causes and fixes.

Software Safety Records Log

The software safety records log is the depository for all the data and evidence that supports the software safety case [ibid.]. It should contain:

- Evidence that the software development plan has been followed
- Results of software safety reviews and audits
- Results of all verification and validation activities
- Analysis of the historic system failure rate, error identification and fixing
- Analysis of the suitability of the development methods and tools.

Software Development Plan

This document is a plan for the computer system's development process and how it is going to be managed over the development life-cycle [ibid.]. This document should include the following information:

- Definition of the project objectives
- Details of the project organisation
- Description of the system development process
- Definition of the intended outputs from the development process
- Project management and control information – schedules and milestones
- Details of the metrics to be recorded during the development process
- Details of the system development environment inc. tools and methods.

The Master Test Plan

This document contains the detailed description of the tests planned for each phase of the system development. It forms the key part of the direct test evidence required for the software safety case, and hence the whole system safety case. This document should contain:

- The identification of each test
- An explanation of the purpose of each test
- Linkage to the system requirement being demonstrated by each test
- A description of the required test environment
- A description of the test conditions and input data to be used
- The acceptance criteria for each test.

It is vital that any safety requirements that require demonstration through test are included in this document, together with an argument as to why the tests chosen are sufficient to demonstrate the satisfaction of these requirements.

Notes

Chillarege 1999, "Software Testing Best Practices", Center for Software Engineering, IBM Research, Technical Report RC 21457, 1999.

Garfinkel, S 2005, "History's Worst Software Bugs", Wired News, Nov 8, 2005.

HSE 1998, "The Use of Computers in Safety Critical Applications" The Health and Safety Executive, Her Majesty's Stationery Office, 1998.

HSE 2001, "Methods for Assessing the Safety Integrity of Safety-Related Software of Uncertain Pedigree (SOUP)" Prepared by Adelard for the Health and Safety Executive, Research report 337/2001, 2001.

IEC 1998-2000, "IEC 61508 (7 parts) Functional Safety of Electrical / Electronic / Programmable Electronic Safety-Related Systems." IEC Geneva, 1998-2000.

Mahon 1981, "Report of the Royal Commission on Inquiry into the Mount Erebus Aircraft Disaster", April 27 1981.

MISRA 1995, "Integrity" MISRA Report 2, Motor Industry Software Reliability Association, Motor Industry Research Association, Nuneaton. February 1995.

MoD 1997, "Requirements for Safety Related Software in Defence Equipment, Part 2: Guidance". Defence Standard 00:55, Issue 2, Defence Procurement Agency, MoD, August 1997.

O'Halloran 1999, "Assessing Safety Critical COTS systems", Towards System Safety – Proceedings of the Seventh Safety-critical Systems Symposium, Huntingdon, UK. Springer 1999.

Chapter Seventeen

Management Factors in Safety Cases

Introduction to Management Factors

Historically, risk and safety analyses started by looking at mechanical equipment failure, over time the analyses have evolved to progress through assessment of electrical components, software systems and now an increasing number of analysts are now attempting to address the risks from humans interacting with the system.

Lagging behind is any attempt to consider the risks posed by management, particularly senior managers, and judging by the results and recommendations, these risks can outweigh those from component failure. The policies and strategies that they make and the cultures created by them, by design or by default predispose accidents to occur – they are just waiting to happen [Redmill 2006].

The management team, board of directors, or whatever they are actually labelled as, has a controlling position over the culture of the organisation and the systems within it – including safety. In many countries, this group is legally obliged to be responsible for the safety systems of the organisation. However, all too often they fall into making incorrect inferences about their safety systems:

1) Nothing has happened *so far* in my system $\Rightarrow$ The system is safe.

This then leads on to a worse inference;

2) The system is safe $\Rightarrow$ Nothing *will* happen in my system.

Statement '2)' is the basic rule of inference of safety engineering, however, the ongoing list of disasters in the world demonstrates that the conclusion of '2)' may never be achieved [Sträter 2005].

The real inference that should be targeted is:

3) The system is safe enough $\Rightarrow$ Nothing intolerable will happen in my system.

Only managers and governors can direct an organisation to achieve and demonstrate this relationship. The organisation may just be a team of six working on a simple project – or it may be a multi-national company employing thousands. The managers may be the whole board of Directors or it may just be the local team-leader. Every level of manager has the responsibility to achieve '3)', if not legally, then at least morally. The major error that comes from believing 1) and 2), is not doing anything to improve the safety system. Although it must be noted that being tolerable does not guarantee freedom from accidents.

The Managers Caused the Accident

A great many of the accidents and disasters around the world that have been used as case studies over the years have cited significant contributions from the management chain e.g. Herald of Free Enterprise, Piper Alpha, Challenger and Bhopal. This text is not going to repeat previous discussions on these topics, rather, it will concentrate on areas where management contributions to accidents happen far more often.

Historical Incidents [BSC 2005-06]

The manager of a UK haulage firm was jailed for manslaughter for three years after one of his drivers had knowingly and with the complicity of the manager, killed a cyclist in August 2003. The driver was also jailed for four years for dangerous driving. Two other persons were each jailed for nine months for helping to destroy tachograph documents (12/05).

A company Director who put profit before safety by ignoring health and safety prohibition notices has been jailed for nine months. A worker fell 12 feet to the ground in an accident after being struck by an excavator machine. During the accident investigation additional safety risks were identified and a prohibition notice was served. This was seen to have been allowed to be continually breached throughout November and December 2004 (01/06).

A construction firm owner who showed 'total contempt' for the safety of his workforce has been jailed for 18 months after a worker was killed when a telehandler toppled over. The employee was working from a 'home-made' basket raised up 25 feet in the air to work on a barn roof in February 2004. The telehandler had defective breaks, missing ballast and had not been serviced for more that two years. The basket had a guardrail that was too low and a gate that opened outwards. The findings also indicated that the firm had not carried out an appropriate risk assessment of the task (03/06).

These incidents are not simply caused by the single person at the 'sharp-end' of these companies. In the first incident, multiple employees of the company all received gaol sentences (this would have effectively wiped the company out). The continual breach of a prohibition notice does not happen because of one worker – it is permitted because the whole management structure allows it to happen. It is the owner of any firm, who is responsible for carrying out risk assessments and preventing un-safe acts. A home-made basket on unserviced equipment is something a management review should reject immediately.

The final statistics make an interesting and tragic summary – three fatalities, 120 months of gaol sentences and all occurring within six months of each other. Out of interest, compare these with the statistics of the three high-fine accidents noted at the end of chapter 12. There were 23 fatalities, £42 million in fines, with six years between them. In the two UK cases, there were no personal convictions for the offences, this has further enhanced the need for reform of the law on corporate responsibility in the UK. The US Chemical Safety and Hazard Investigation Board are still investigating the Texas City explosion.

Managers and the Law

NOTE: I am a professional engineer, not a lawyer, and this text has not been prepared by a lawyer, so if you have any *legal* doubt about the content of your safety case, please consult one.

Many of the prosecutions that come to court are seeking to punish negligence, either by an individual or by a corporation as a whole. In countries which have a legal basis built on the concept of common law, the law tests four concepts for negligence to be proven: Causation, Foreseeability, Preventability and Reasonableness [Anderson 2006].

> Causation: This test asks whether harm could occur because of the unsafe matter on which the charge of negligence could be based.
>
> Foreseeability: Did you know? Or ought you have known? about a potential source of physical injury or damage to the health of people.
>
> Preventability: This question asks whether there is a practicable way or alternative to how things would be done, which would better respond to a hazard actually happening.
>
> Reasonableness: This requires a judgement as to the balance of the significance of the risk vs. the effort required to reduce the risk to an acceptable level.

All of these tests can be answered by a well produced safety case. The causation can be answered by the hazard identification, the foreseeability can be answered by the risk and safety analysis, the preventability and reasonableness can be answered by the ALARP analysis and justification.

An alternative description comes based on the 'Duty of Care' concept, where the components of a negligence cause of action are considered as follows;

- Duty: A person owes a duty of care to another when a reasonable person would foresee that there would be an exposure to risk, as judged at the time of the events, not with hindsight.

- Breach of Duty: A subjective and objective judgement on whether there was a decision to continue exposing the person to risk.

- Causation: It must be shown that the particular acts or omissions were the cause of the loss or damage sustained.

- Damage: There must have been some loss or damage flowing naturally from the breach of the duty of care. This may be physical, economic or reputational.

Within the UK as this book is being written, a consultation is underway on the concept of corporate manslaughter. The current common law test to impose criminal responsibility on a company can only arise where there is a 'controlling mind' whose actions and intentions can be imputed to the company. This is where a person is in control of the company's affairs to such a degree that the company can be said to think and act through them. This is tested by reference to the detailed work patterns of the manager, and not the job title or description given to that person, as this is considered irrelevant in law. However, in most large organisations, there is often no single person who acts as a 'controlling mind', and many issues of health and safety are often delegated to junior managers who do not pass the 'controlling minds' test.

There is growing concern over the idea of criminal culpability based on individual responsibility and the increasing recognition of the potential for widespread harm from corporate activity. Draft legislation has been laid before the UK parliament to bring in a new offence of corporate manslaughter. The new test is proposed to be 'if a company has been managed or organised by its senior managers so as to cause a person's death, and if it amounted to a *gross breach* of the duty of care owed to the deceased'. The organisation will be treated as a fictitious person and as such could not go to gaol, however the intent appears to be to have an unlimited fine available. This could effectively close down any organisation by matching the level of the fine to the value of their assets.

In other countries e.g. Scotland and Canada, the corporate manslaughter legislation is seeking to identify individuals for criminal prosecution, with prison terns as the punishment.

Promoting a Safety Culture

There is no standard definition for a 'culture' with respect to safety. The word 'culture' has several dictionary definitions including [Longman 1987]:

- The development of the mind
- Enlightenment and excellence of taste
- The customary beliefs and social forms of a racial, religious or social group.

But my personal favourite from the same reference and most applicable in this chapter is:

> The socially transmitted pattern of human behaviour that includes thought, speech, action, institutions and artefacts.

When given the goal of maximising safety and health, these defined components of a culture can be very powerful indeed. It has been stated that the ideal safety culture is the 'engine' that continues to propel the system towards its safety goal, *regardless* of the characteristics of the leadership and the contemporary commercial concerns [Reason 1997]. This is an important concept, the amount of independence there is from the Board and the Accountants. In a sense, a safety culture cannot be achieved without the guidance of senior leaders, and the financial commitment to cover the required resource, but a safety culture can also be ruined by inappropriate actions from both areas.

Four critical sub-components of a safety culture have been identified [ibid.], and they are worthy of summary again. They are a reporting culture, a just culture, a flexible culture and an informed culture.

A Reporting Culture

Five factors have been identified as assisting in developing a reporting culture, where useful (safety case evidence) information is recorded and made available to be assessed such that a more serious event can be mitigated. These are:

- Indemnity against discipline – so far as is reasonably practicable
- Confidentiality and/or de-identification
- The separation of the recording body from the area responsible for discipline
- Rapid, useful, accessible and intelligible feedback from the process
- Ease of making and submitting a report.

A Just Culture

In this case 'just' means justice, if someone has caused an incident through recklessness, negligence or malevolent behaviour, justice must be done and seen to be done. However, the vast majority of unsafe acts are caused through situational and systemic factors, for which the apportionment of blame on an individual is not appropriate and a whole team may need some attention. The justness of the culture should seek to match the crime with a suitable punishment. In the UK, the punishment for reckless or dangerous driving may be a ban from driving for a period of months or years; the most severe penalty (as viewed by the driver) is having to retake and pass the driving test to regain your licence to drive. The punishment of re-training is sometimes very appropriate. At some point though, there will be a case where dismissal is required, perhaps for gross negligence or perhaps repeat 'offenders'. The company code for safety offences should be explicit, published and agreed by all staff.

A Flexible Culture

A flexible culture is one where the capability provided by the staff and leaders is agile in nature. The concept is one of 'best leader for best situation', allowing and encouraging this to happen, and thriving on it. To create this type of culture, an organisation must invest in the quality and training of key leading personnel so that they can direct and lead safe working practices without needing to refer to standard procedures or rigid hierarchical rule-sets – they *feel* what is safe enough. When this happens, people are not *taking* risks, they are *running* risks [Reason 1997].

A Learning Culture

In this culture, the organisation discovers features of itself, some it will like and promote, others will not be liked and should be removed. Both sets of features have to be acted upon – the promotion and removal are verbs – doing words, so actions must be developed in order to 'do'. The actions must have initiators, how bad must something be before it is removed? And the initiators must be looked for and understood. The process of searching, finding, understanding and correcting is a pretty good definition of learning in this case.

Summary of Safety Culture Governance

Governing is different from managing, activities are managed for the sake of the activity, activities are governed for the sake of the organisation, or in the domain of safety, for the sake of the culture of the organisation. When there is an absence of governance, the underlying business activities tend to drift away from that which is most desirable for the business. The business will also find it hard to adapt to change and to control its risk profile [Vickers 2006]. Senior managers must be able to influence the way in which an organisation implements its safety policies and they can do this through governance following five principles [ibid.]:

1. You can't govern everything, govern the critical areas – safety is critical
2. Set system objectives for safety e.g. reducing mishaps or down/loss time
3. Set a key performance indicator for each objective – what is good/better/bad?
4. Ensure systems exist to measure the key performance indicators
5. Regularly review the performance indicator data.

The way the leaders and management team applies safety principles to the workplace, product or process directly affects the credibility and integrity of the safety case that is produced. It should be sung and shouted about if it is good – if it isn't good, the managers should re-read this chapter again.

Evidence from Managers for the Safety Case

Managers have an essential part to play in constructing a safety case and writing the safety case report. Whilst it is not critical for them to have day-to-day involvement, there does need to be regular contact between the controlling minds of the organisation and the products and services coming out of it. I would at least expect the person with delegated responsibility for safety to have read and signed all the safety cases produced in the organisation. Otherwise, how does s/he know how much corporate risk the organisation is exposed to?

Much of the safety case is directly related to the system of interest – what hazards it has and what effects can those hazards produce. However, a significant amount of evidence in a safety case comes from corporate processes, its experiences and the knowledge and ability of its staff. One phrase that is being discussed these days is the value of corporate memory, which combines most of the above and is often released, retired or made redundant far too easily, but don't get me started there!

The management chain should have a corporate safety policy, this should dictate how safety is to be managed through all corporate affairs and all projects undertaken by the organisation. This is the top piece of evidence as this sets and defines the rigour and fidelity of all safety reporting work. From the safety policy should flow safety management plans. Often, these are written by project staff with respect to the management of the project alone. There should be significant effort from the management levels in the organisation into understanding and contributing to the safety management plan, since these are the main mechanism for controlling personnel and the company safety-risk exposure. If this is not clear in any particular organisation, how can the project safety management plan be consistent and appropriate? The safety management plan should identify who in the organisation is responsible for safety, what analyses and work is going to be done, how it is going to be checked, by whom and by when. Each step in the safety management chain should have their own safety management plan, so that they can demonstrate that they have not been negligent in discharging their duty of care. These intermediate safety plans can then all be included or cited directly in any safety case that is produced by the organisation – all as evidence of care and safe management practices.

Within a safety case or safety report, there should be a single-sheet statement of contemporary residual risk that summarises the level of risk posed by the system or equipment of interest. The safety engineers working on the safety case will usually formulate this statement at critical project milestones or gateways. It is recommended that the management chain be involved in signing this piece of paper. This acts in two ways, firstly it acts as an endorsement of the safety work that has been completed up to the publishing of the report, and second it forces the management chain to recognise and accept (on behalf of the company) that there really is some residual risk. The management chain could, if it was needed, sum up all the contemporary residual risks from all the projects, so that at any given time, the company is fully aware of its total risk portfolio.

Notes

Anderson 2006, "Common Law Safety Case Approaches to Safety Critical Systems Assurance", Developments in Risk-based Approaches to Safety, proceedings of the fourteenth safety-critical systems symposium, Bristol, February 2006.

BSC 2005-06, "Safety Management", The British Safety Council, London 2006.

Reason 1997, "Managing the Risks of Organisational Accidents", Ashgate Publishing Limited, Aldershot, 1997.

Redmill 2006, "Understanding the Risks Posed by Management", Developments in Risk-based Approaches to Safety, proceedings of the fourteenth safety-critical systems symposium, Bristol, February 2006.

Strater 2005, "Cognition and Safety – An Integrated Approach to System Design and Assessment", Ashgate Publishing Limited, Aldershot, 2005.

Vickers 2006, "Governing Safety Management", Developments in Risk-based Approaches to Safety, proceedings of the fourteenth safety-critical systems symposium, Bristol, February 2006.

Chapter Eighteen

Independent Safety Review

The Principles of a Review

The two main objectives of having any kind of review are firstly, to check that the work undertaken to date is correct enough to be of value for the purpose. Secondly, to get some feedback and advice from the reviewer with pointers as to what has been good and what needs more attention. I was once involved in a personal motivation exercise where two people from the group were asked to leave the room, the remainder were told that when each returned they would be given an object to pass on to an unidentified individual. The first volunteer would be 'coached' by the room-based group by being told 'no' when they didn't do the correct thing. The second volunteer would be 'coached' by the room-based group by being told 'no' when they did the wrong thing, but also 'yes' when they made a correct action. When the first volunteer came back, they had no idea what they were supposed to do, received only negative comments, quickly realised the situation and just sat at the front of the room. The second volunteer also had no idea what they were supposed to do, but as they nervously bobbed about at the front, they picked up the positive signs when they 'bobbed' towards the unidentified person they were to give the object to. It took about four minutes. This is an example of real-time review in action and actually working, everyone took important lessons from this, and I suggest you try it sometime.

How Independent is 'Independent'?

There still is a great deal of debate about how independent an independent review has to be, and there are two opposing factors to consider in this discussion. Firstly, the reviewer must be independent enough to be able to provide an expert, professional opinion without vulnerability to commercial, project or other corporate pressures [Froome 2005]. Secondly, the reviewer must not be so independent as to be essentially unaware of the details of the project systems and their operation. Levels of required independence may be set for levels of project risk profile, project size or project sensitivity. At the lowest level of independence is the co-worker from the same team. The levels then progress through a scale up to personnel having different 'controlling-mind' directors and even being from an independent company altogether, although this last level can give rise to commercial difficulties. The company receiving the audit should not have to accept a reviewer from a competitor organisation, it is far better for a mutually acceptable reviewer to be chosen at the start of a project, rather than have difficult arguments when it comes to the first review or audit.

On one of the projects I have worked on, the project safety manager said he would rather have an experienced view point than an independent, less experienced one. As recorded by the IEE/BCS independent safety assurance working group [Smith 2005], there is not a simple definition of independence that fits, so it is probably better to take the view that 'independence is in the eye of the beholder'.

A Review by the Regulators

In the US, the OSHA carries out inspections and reviews of organisations under its jurisdiction, in the UK the HSE has the same role. In many cases there are laws that dictate that the regulators are required to review an organisation's safety case or safety plan prior to it being declared acceptable. This review is wide ranging covering not only the safety document involved, but all of the supporting evidence, tests and engineering judgement used to construct the whole safety statement. This is why it is vital to record all the evidence and to keep it available in case the regulators ask for it. There usually has to be a special reason for them to ask for *all* the supporting information e.g. an accident or perhaps a change in the legislation.

The reviewers can also note safety concerns outwith the scope of the specific safety case they are dealing with, and can impose penalties, requirement notices (requiring specific work to be done) or even prohibition notices, where all work must stop.

In the US, the Occupational Safety and Health Review Commission (OSHRC) is the independent Federal agency created to decide contests of citations or penalties resulting from OSHA inspections of American work places. The Review Commission, therefore, functions as an administrative court, with established procedures for conducting hearings, receiving evidence and rendering decisions by its administrative law judges.

In the UK rail sector, following recommendations by Lord Cullen in his Ladbroke Grove Rail Inquiry Part 2 Report, a number of amendments were introduced to the Railway Safety Case Regulations on 1 April 2003 [HSE 2003]. These simplify the safety case assessment process, and require each railway operator to obtain an annual health and safety audit from an *independent* competent body. The regulations still require railway operators to have a 'safety case' formally accepted by the HSE. In this instance, the safety case is a comprehensive core safety document setting out the operator's policy, objectives, organisation and management system for health and safety, and the risk controls that are in place.

Assessor, Advisor or Auditor

In many industries around the world the independent safety review is carried out by a person or team known by the three-letter-acronym, the ISA. The 'I' part (Independent) has been discussed above, the 'S' part is simply 'safety', it is the 'A' part that provokes the most discussion. The role and scope of the person or team that is the ISA is very different if they are assessors, advisors or auditors. It is vitally important to get this function agreed at the early stages of a project, rather than having a difficult discussion about it when the first review is coming up, or worse actually underway. Fortunately there is some guidance as to what each word means and implies.

Advisor

There is a judgement call made by ISAs about how much specific advice can be given whilst still maintaining independence, and not being in a position of writing the safety documents themselves and hence owning the argument rather than the correct duty holder. However, the ISA (advisor) role may provide generic advice on the acceptability or otherwise of some proposed safety argument, but it is not very helpful if the ISA maintains this without giving reasons [Froome 2005]. One possible criterion to follow is that the advice can be given when it is not specific to the project, but still supports the project's decision making. The ISA can illustrate how modifications and revisions should be performed by reference to published standard guidance or papers, or through comparison with other similar projects and good practice [ibid.].

Auditor

This role is far clearer in its scope as there is already a common acceptance of what auditors do in the financial and quality sectors. The auditor will check for conformance to standards, legislation and good practice. Within the safety domain, the ISA (auditor) will want to check that the safety requirements identified are complete, correct, consistent and achievable with respect to the resources allocated. Usually, there is a requirement for the auditor to check that the project's safety tasks are being conducted in accordance with internal documents as well e.g. the organisation's safety policy and the project's safety management plan.

A compliance matrix is often used, where each requirement of a standard or plan is explicitly recorded and demonstration evidence is identified or linked to the particular clause in question. A positive working culture can be established if this is already done as part of the safety document set, so that the auditor can easily check this and follow the links.

The auditor is unlikely to have the scope to actually check the soundness of the reasoned safety argument, an audit would typically check report sections, evidence items and headings for content, not whether the overall safety arguments being put forward were actually valid [Smith 2005].

Assessor

The ISA (assessor) role is again slightly different from the auditor and advisor roles. The assessor goes further than the auditor, usually having the scope to review the evidence, arguments and reasons being put forward as well as the review of compliance with internal and external standards. The IEE/BCS working group has developed a definition for this specific role as follows [ibid.]:

> Independent safety assessment is the formation of a judgement, separate and independent from any system design, development or operational personnel, that the safety requirements for the system are appropriate and adequate for the planned application and that the system satisfies those safety requirements.

Now whilst a great deal of this definition deals with the concern of independence, the assessor part may be summarised as the formation of a judgement as to whether there is a decent set of safety requirements, and whether the system has met them. This is perfectly acceptable, but it doesn't give much indication about how the assessor is to interface with the project. There is no discussion indicated, the assessor will form a judgement … and that's it. The assessor should have the scope to take more of an active role. Safety assessments (and audits for that matter) should not only be used to find weaknesses and non-conformances; they should also serve to build strength and confidence in the safety management process. They should develop, encourage and spread good practices between safety cases and safety teams.

Competency of the Reviewer

It is widely recognised that there is no formal definition of what qualifications or experience should be expected of an ISA (any 'A') [Smith 2005]. Produced in collaboration with the British Computer Society and the Health & Safety Executive, a new set of competency guidelines have been published to help companies assess and record the competencies of staff working in safety-related applications including reviewers. The guidelines contain a set of competency statements and guidance on an assessment procedure. The document may be used, entirely voluntarily, by organisations to support the assessment of competency of personnel. Specific competencies identified in this document are of four types and are spread across 12 safety functions including hazard and risk assessment and independent safety assessment [IEE 2006]:

> Technical skills; for example, hazard analysis, report writing.
>
> Behavioural skills; for example, personal integrity, interpersonal skills, problem solving, attention to detail.
>
> Underpinning knowledge; for example, a person performing a hazard identification must have knowledge of the particular application to be able to identify the likely hazards that exist.
>
> Underpinning understanding; for example, it is unlikely that somebody could establish risk tolerability levels for a particular problem without an understanding of the principles of safety and risk.

The initial scope of the guidelines was defined as competencies for mainstream, safety-related practitioners working on Electrical, Electronic and Programmable

Electronic Systems (E/E/PES) as defined by the international generic standard IEC 61508 on functional safety. As this standard is not part of UK Law, the advice is only guidance and not mandatory.

The Terms of Reference

The terms of reference (ToR) for an ISA (any 'A') should be stated in the safety management plan. The main stakeholders, including the ISAs themselves, should agree this. The scope should list the standards and legislation applicable to the system, operation or product in question and identify a timing plan for when audits shall be performed. The reporting procedures and formats should also be stated, these are likely to be made contract deliverables and may certainly be used as evidence in the safety case. The ToR may even list out the individual safety documents that are highlighted for specific attention.

Often there will be guidance in industry standards for the role and terms of reference for the ISA (usually a specific 'A' but variable over different industries). The following is an extract from a safety case I was involved with, and it defined the way the ISA was going to fit in with the safety tasks of the project.

The ISA shall be responsible for:
- Reviewing and advising the customer project manager regarding the acceptability of the safety requirement as defined by the Safety Case.
- Auditing the project for compliance with the standards, guidelines or codes of practice in accordance with an agreed audit plan.
- Auditing the safety deliverables produced by the contractor, and ensuring that work is performed in accordance with the contracted standard in accordance with agreed project plans (Safety Management Plan, Human Factors Integration Plan, Software Development Plan).
- Reporting on findings equally to the customer and the contractor.

In projects which extend over more than, say a year, or ones that are particularly complex, the ISA team should be invited to produce and follow an audit plan, rather than rely on a small section of text within the safety case itself. This way, a specific document can be published for all the stakeholders to review, and the publication cycle doesn't need to get bogged down in the full review cycle of a safety case and all its associated documents.

Notes

Froome 2005, "Independent Safety Assessment of Safety Arguments", Adelard LLP, Constituents of Modern System-safety Thinking – proceedings of the thirteenth safety-critical systems symposium, Southampton, February 2005.

HSE 2003, "Railways (Safety Case) Regulations 2000 including 2001 and 2003 amendments" Health and Safety Executive Ref. L52, HMSO, 2003.

IEE 2006, "Précis of Safety, Competency and Commitment: Competency", IEE (now subsumed into the IET), London 2006.

Smith 2005, "The IEE/BCS Independent Safety Assurance Working Group", Frazer-Nash, Constituents of Modern System-safety Thinking – proceedings of the thirteenth safety-critical systems symposium, Southampton, February 2005.

Chapter Nineteen

Presentation of the Safety Case

Introduction to Presenting Safety Cases

Ultimately, the construction of a safety case is dedicated to communicating about the safety property of some system, process or item of equipment. In order to do this, the safety case must be able to be presented to a manager, reviewer, regulator or other interested stakeholder. Since the inception of reporting about safety, the presentation has been done using the good ol' paper report, often with a pre-specified contents and structure. With the development of widely available IT power and functionality, there is an alternative presentation medium available – the computer. There are advantages and disadvantages in using both mediums, but there is also plenty of good advice and commentary on how to get the best out of the chosen medium and present the best case for safety that is possible.

The Paper-based Safety Case

Many safety cases and safety reports are written specifically for a paper-based presentation. This is probably done more by implicit habit than intentional design, most private and business communication that needs to be recorded is done so on paper. Humans are experienced at writing for paper – ever since school and college days, people have been taught how to write for a paper medium.

There is absolutely nothing wrong with this as conversely, ever since those classroom days, people have also been taught to read from a paper medium! If you want someone to understand the communication, it is best to have it in a format that can be readily assimilated. Paper works just fine.

The advantages of using a paper format include as follows:

- The medium is readily available and cheap
- Word-processing tools are readily available and cheap
- Wizards and templates can (maybe!) help in report design
- The information is portable and distributable
- It is a familiar medium to both writer and reader
- A report is readily accepted as contemporary with the publication date
- Properly referenced and signed it is admissible in Law
- Graphics and text may be readily combined
- Image and importance can be demonstrated through coverings.

The disadvantages of using a paper-based format include as follows:

- Paper and printing has a certain environmental impact
- Complex discussions are difficult to represent in hard copy
- The formatting of a paper-based report can take longer than writing the text
- Paper-based reports are numerous and unlikely to stand out
- Paper-based reports can get lost or damaged
- Indexing and searching have to be performed linearly
- There can be extensive update and re-print costs
- Wide variability in text style and writing techniques.

Recommended Layouts for a Paper-based Safety Case

There are a multitude of recommended layouts for a safety case depending on where you are and what industry you are working in. A few examples will be given here for comparison and contrast purposes. If you are obliged to adhere to a specific standard, you had better stick to that. If there is no standard, by all means use one of these or a combination that best suits your presentation needs.

As Recommended for UK Defence Safety Cases [MoD 2004]

- Executive summary. This should enable the Duty Holder to provide assurance to stakeholders that he/she is content with the progression of work and that safety requirements have been, or will be met.
- Summary of system description. A brief description should be given noting that a full description is contained within the safety case [as a document suite].
- Assumptions. The assumptions that underpin the scope of the safety case, or the safety requirements, arguments or evidence should be stated.
- Progress against the programme. An indication of the current status relative to expectations within the programme, and progress on safety management since the previous safety report.
- Meeting safety requirements. This section should include a description of the principal, agreed safety requirements (e.g. ALARP), and a summary of the argument and evidence that demonstrates how the safety requirements have been, or will be met. A statement about the contemporary residual risk should be made.
- Emergency/Contingency arrangements. A statement confirming that appropriate arrangements have been or will be put in place and identification of any areas where such arrangements are likely to be inadequate.
- Operational information. This section should contain the output from the safety case that is relevant to the management of operational safety, including, the main risk areas and any limitations of use or operational capability.
- Independent safety auditor's report. Where an ISA is engaged, they should prepare a formal report for inclusion in the safety case report.
- Conclusions and recommendations. This should include an overall assessment of the safety of the system and any recommendations to enable any issues to be resolved.
- References. A list of key reference documents should be provided [including key test and process evidence, hazard logs, software and human integration reports, and any other evidence being used to support the safety case].

As Recommended for European Civil Aviation [EUROCONTROL 2005 & Fowler and Tiemeyer 2006]

- Executive summary. This should provide the reader with an overview of what the safety case is about, what it is trying to show and for whom, a summary of the conclusions, caveats and recommendations.
- Introduction. The introduction should include the historical background to the safety case; a statement of the aim and purpose of the report; the scope and boundary of the safety case and the purpose of each section of the document.
- System description. This should provide a description of the system to which the safety case applies, including its operational environment, interfaces and boundaries.
- Overall safety argument. This section should describe and explain the highest levels of the safety argument structure including the main safety claim, a definition of what is meant by 'safe' and a description of the operational context of the system.
- Safety Argument and Evidence. These areas should present each of the principle safety arguments in turn together with the supporting evidence which shows that each of the arguments is valid.
- Assumptions. All the assumptions on which the safety case depends should be presented directly and/or by reference. Each assumption must be shown to be valid or at least reasonable according to the circumstances.
- Issues. Any outstanding safety issues that must be resolved before the safety claim can be considered to be valid. The responsibilities and timescales for resolution should also be listed.
- Limitations. Any limitations or restrictions that need to be placed on the deployment and/or operation of the system should be stated and explained.
- Conclusions. The main conclusion should refer to the original claim, reassert its validity and note any caveats.
- Recommendations. Recommendations [for additional work etc.] are not mandatory [orders], so ones on the actual use of the safety case are most appropriate. Recommendations must not contain any statements that might undermine the conclusions.

From the USA

As the US doesn't explicitly use the 'safety case' phrase, there are no readily available US recommendations for its layout. However the Occupational Safety and Health administration does give advice on the construct of a safety and health programme, including the completion of a 'safety and health program report'. This advice [OSHA 1989] has been used to develop the following list, for full details of the requirements, please see the reference.

Part 1: Management Leadership and Employee Involvement
- Worksite safety policy
- Current safety goals and objectives
- Task and system description
- Orientation outline for staff, visitors and contractors
- Evaluation of safety and health responsibilities
- Budget showing money allocated to safety and health
- Evidence of employee involvement.

Part 2: Worksite Analysis
- Results of baseline site hazard survey with notation for hazard correction
- Employee reports of hazards
- Historical mishap investigation reports
- Trend analysis results
- Procedures for change analysis, which include hazard considerations.

Part 3: Hazard Prevention and Control
- Maintenance records
- Preventative maintenance procedures
- Site safety and health rules
- Emergency drill procedures
- Health surveillance and monitoring procedures
- Reports, investigations and corrective actions taken for near misses
- Specific OSHA mandated procedures for specific hazards.

Part 4: Training
- Program of yearly training topics
- Employee training recording procedures and data.

It is not my intention to critique these layouts or contents, a comparison between them should suffice to indicate the areas to consider when constructing a safety case document.

All three approaches have a system description near the start of the report; all also contain details of hazard and safety analysis. The MoD and OSHA both have explicit sections on emergency procedures; the MoD and the EUROCONTROL recommendations both give a great deal of effort to recording the specific safety assumptions. The MoD and OSHA guidance do have an acknowledgement of the idea that there is an on-going process in place – the MoD explicitly asks for progress against the programme, the OSHA asks for trend analysis and a current safety goal ('current' implying that this changes over time). The OSHA list stands out by specifically asking for a declaration of the budget allocated to safety and health. It is difficult to see this happening in the UK or EUROPE, although as part of ALARP there is a cost-benefit aspect to the accompanying analysis, so perhaps that is not so far away.

The training and health surveillance aspects are made more explicit in the US list, this is due to the combined scope of safety *and health* that is not present in the MoD and EUROCONTROL lists. The health surveillance that does occur in these two lists is the health of the system being assessed. Finally, the MoD and the EUROCONTROL lists explicitly ask for the definition of operational limitations and any safety issues that are still outstanding. Overall, the three approaches do have a great deal of overlap in content, as well as several areas of bespoke content that are individual to the industry or country.

Having now seen the significantly large content requirements for each of these approaches, it is not difficult to see how these documents can quickly become extensive in size and complexity. This can make them very difficult to manage and maintain over time, as required in many safety case approaches.

The IT-based Safety Case

With the advent of powerful computer technology and additional information technology tools and processes, it is now possible to present a safety case in a totally electronic format. This goes much further than just being a paper document in an electronic file format, even with the template capability and section to reference hyper-linking available in many contemporary word processing packages. In these situations, the safety case is still constructed and maintained with paper in mind as a presentation and recording format.

A digital system for constructing a safety case is now possible, where recording evidence and safety arguments can be done wholly within the IT system. Presentation is done using the graphical-user-interface (GUI) of the system using dedicated notations, and the availability of external hyper-linking using shared file directories and even the (secure functionality of the) internet. Together, they can provide direct access to all the required standards, test reports, and safety related reference documents from internal and other system domains.

Advantages of using an IT-based system include as follows:

- Ability to utilise all functionality of the IT system
- Allows direct hyper-linking to all required information and evidence
- IT systems are now in place at all worksites
- Updating and issue can be done on a one-to-many posting process
- Full graphics and multimedia capability supported
- File and data storage can be more efficient
- More complex safety arguments can be supported
- Standard notational styles may be used
- Can still print out on paper if required
- Allows more complex searching tasks.

The disadvantages of using an IT-based system include the following:

- All users have to have similar IT systems
- All current and future users have to be trained to use the IT system
- Lack of competence in the tool can erode confidence in the safety case
- All readers have to understand how to read/use IT systems
- Frailties of viruses and software bugs in IT systems can cause data loss
- Dedicated tools can be expensive
- Difficulty in introducing new systems into large organisations
- Tailoring to existing organisational practices requires specialist knowledge
- Levels of commercial security may prohibit internal/external hyper-linking

- Hyper-links may fail and cause loss of configuration control
- Some standards required *all* tool use to be subjected to the safety case.

Recommended Layouts for an IT-based Safety Case

There are no specific standards that give advice on how to present a safety case in digital form, so if you can derive a bespoke one, by all means carry on but bear in mind the disadvantages noted above.

Essentially the main elements of a safety case are:

- Claims about a property of the system or subsystem
- Arguments providing reasoning and inference about the way that a claim has been satisfied
- Evidence providing the data and information to support the argument chain
- Contexts about the operating environment, standards and usage assumptions.

These elements enable a retrospective style of safety case to be constructed, where a claim is being made about an existing or developing system. Alternatively, the safety case can be composed in a style that is more predictive in nature, where safety goals have been set for a future or developing system. In the predictive case, 'strategies' and target 'objectives' are used in the structures. There is no special rule that says you must use one or other style in a particular circumstance, although customers or regulators may express strong preferences. Indeed it is entirely feasible to combine both styles as the safety case planning and reporting progress through time. For example, the objectives in the plan could become the items of evidence in the safety report. Similarly, strategies develop into arguments and goals can become the safety claims.

A specific graphical display system has been put forward to enable the presentation of a safety case organised in these styles, it is known as 'Goal Structuring Notation' or GSN [Kelly 2003]. The principle elements of GSN are shown in Figure 19.1 and 19.2. These elements may then be joined together using links that show how the whole structure combines together to form a cohesive argument and justification for the demonstration of the primary goal. The links can be given directional arrows to indicate whether the structure is a top-down style or a bottom-up style.

These two styles of representation can be used to show a decomposition and re-composition concept allowing a synergy to be developed with the system engineering 'V' diagram. In this instance, the decomposition left-side down the 'V' can utilise the predictive goal based schema, and the re-composition right-side of the 'V' can utilise the more retrospective claim-based schema.

The notation is infinitely flexible and can be used to represent any complex argument, not just safety based. It is particularly suited to the types of standards and regulations that are goal-based rather than prescriptive in nature, so any industry could make use of them. The flexibility means that it is highly likely that completely different goal structures could be drawn to satisfy the same complex problem – and both would be correct, just from a different point of view. This is one of the problems with these notation-based safety arguments, there is no formal reference or standard to check against. Some work is being done in this area in terms of defining safety argument patterns that can be re-used in similar situations, this gains the risks and benefits of producing a series of black-box COTS safety arguments, that only a specialist few can potentially understand and interpret.

A typical GSN structure may be composed as shown in figure 19.2. There are no absolute rules governing which element should follow which element, the only rules in this area are that there should be a goal at the top and evidence items at the bottom. It is recommended that the area in between the top goal and evidence should achieve clarity and understanding. A strategy element should always be used to explain how proposed sub-goals or solutions satisfy the higher goal, unless this will be obvious to all readers of the safety case report. Frequently networks are developed as plans mature and the safety case develops, that shown in figure 19.2 is an interim network in the course of development.

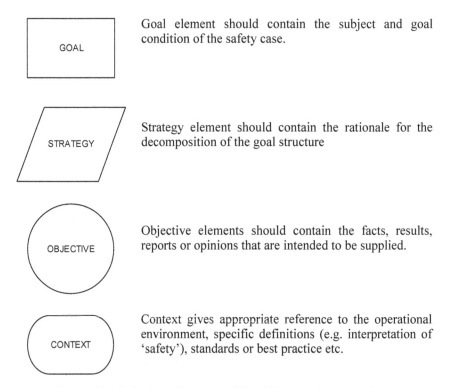

Goal element should contain the subject and goal condition of the safety case.

Strategy element should contain the rationale for the decomposition of the goal structure

Objective elements should contain the facts, results, reports or opinions that are intended to be supplied.

Context gives appropriate reference to the operational environment, specific definitions (e.g. interpretation of 'safety'), standards or best practice etc.

Figure 19.1 Principle Elements of Goal Structuring Notation

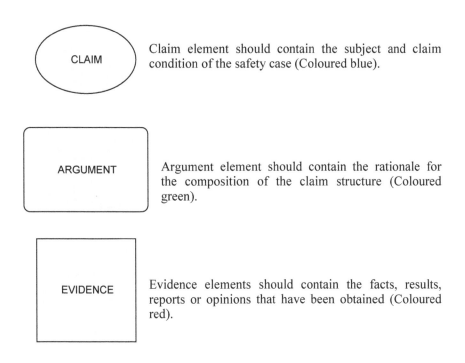

Claim element should contain the subject and claim condition of the safety case (Coloured blue).

Argument element should contain the rationale for the composition of the claim structure (Coloured green).

Evidence elements should contain the facts, results, reports or opinions that have been obtained (Coloured red).

Figure 19.2 Principle Elements of Claim Arguing Notation

Goal Structuring Notation Tool Support

There are a few tools available to assist in the presentation of an IT-based safety case. Some of these tools e.g. ASCE give very powerful functionality when composing the safety case, for example [Adelard 2006];

- Information mapping that supports the generation and updating of reports including traceability of changes in core documents.
- Summarising, reporting and tracking information held in other documents including hazard logs, requirements management tools, or other information sources such as Excel, Access, Oracle, or Word.
- A plug-in architecture providing a unique user extensible capability - create bespoke applications using VBScript.
- Track the impact of changes and propagate the impact visually across the network.
- Support for an XML standard format for assurance and safety cases, this enables the exchange of case information between co-operating tools.
- Tools to support the construction and maintenance of safety case reports, including validation of information maps and hyperlinks.

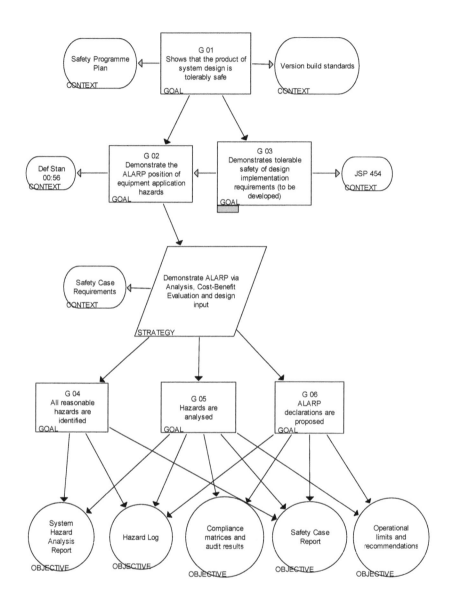

Figure 19.3 Example Goal Structuring Notation Structure

Notes

Adelard 2006, "The Assurance and Safety Case Environment – ASCE", Adelard LLP, City University, London, 2006.

EUROCONTROL 2005 & Fowler and Tiemeyer 2006, "Safety Case Development – A Practical Guide", Developments in Risk-based Approaches to Safety, proceedings of the fourteenth safety-critical systems symposium, Bristol, February 2006.

Kelly 2003, "A Systematic Approach to Safety Case Management", University of York, York, UK, 2003.

MoD 2004: "Safety Management Requirements for Defence Systems Part 2" Interim Defence Standard 00:56, Issue 3. Ministry of Defence, December 2004

OSHA 1989, "Safety and Health Program Management Guidelines" US Department of Labor, Occupational Safety and Health Administration, 1989.

Chapter Twenty

Maintenance of the Safety Case

What Happens to Safety Cases?

There are generally two scenarios for safety cases – they are either used or they are not. When a safety case is not used, the safety case argument will remain valid for a short while – perhaps a year or so, but then critical staff will have turned over, new products or procedures will be brought into the system, the product or equipment will be used in a different way, or there may even have been a serious incident which was not originally envisaged. Eventually, the real system will have diverged so far from that represented by the safety case that the safety case is no longer valid or useful. The lack of an appropriate safety case, hazard analysis or risk assessment is regularly used in legal prosecutions as demonstration of negligence. This is absolutely correct, as having out of date information is actually worse than not having any information at all. With nothing to rely on, there is a belief that things are 'risky', if all the risk analysis was done a few years ago there is a belief that nothing has changed so the safety case still applies. This is extremely unlikely, there will usually have been multiple minor modifications and changes, which on their own may appear insignificant, but together could add up to a major step-change in operational use.

There are several areas, which deserve special attention where change can catch out the best of safety cases: ALARP arguments, operational use and improvement, legislation shift and mishap occurrence in a related field. Each of these will now be considered in turn, although there is no priority implied by the order.

ALARP Arguments

If the safety case is predicated on an argument based on risk exposure being As Low As Reasonably Practicable (ALARP), a judgement will have been made about two things – the risk profile exhibited by the system and, the resources of time, trouble and expense in reducing that risk further. Over time, the perception of both areas can change, the risk profile might be viewed with more dread by the public as more research information is published; the time, trouble and expense factors may change with technological development. What was once thought expensively prohibitive (e.g. use of multiple remote control robots to sense for poisonous atmospheres), might become significantly cheaper over time. The development of new ceramics and polymers may provide improved personal protective equipment, so that certain tasks become more possible from a risk perspective.

However, if the ALARP judgements and the assumptions they are founded upon were not reviewed, how would anyone ever know?

The decision point for the cost-benefit analysis does change with time. The value of a prevented fatality, which is a fundamental criterion within ALARP justifications, does increase over time. It does this in accordance with national inflation or when there is a re-calculation of some of the factors used to determine the value (see chapter 12). What was a marginal case 'n' years ago, may now be one where further action is required.

Operational Use and Improvement

The system operators and original system designers are often separated by at least 5 years, sometimes many more. The poor old engineers who were undertaking the original design and constructing the first safety case had to make educated guesses as to how the system was going to be used in the field. In areas where this was difficult, operational limitations were imposed or recommended. Many end-users do not hear about the limitations, or simply do not read the full details in the instruction manual (who does?). Some piece of equipment may be perfectly acceptable in a mild, temperate environment, but take it to the desert or the frozen wastes, and the dust or ice may render critical components useless.

These might be obvious, but one anecdotal story I heard was of a flight guidance computer, that was told that the 'North Pole' would always be 'up'. This worked fine, until the trial aircraft went over the equator.

Of course other operational factors change as well, in the oil and gas industry as specific oil fields run low, the compositional mix of the raw product changes – it contains more sediment and more water. This dramatically effects the mass flow rate through the installation pipework. One of the effects is that the pipework can then start to suffer resonant vibrations that were not there when production originally began.

If the operational use and environment change significantly over time, even just through continuous improvement, then the assumed states in the safety case and boundaries for the hazard analysis will be incorrect. Key hazards will have been ignored, these could now show themselves and bite you very hard.

Legislation Shift

As society tends to become more risk averse, there is pressure to increase legislation to protect society against industrial risk, or to protect the environment from industrial waste. This may not necessarily lead to a fatal flaw in a safety argument, but it may mean that coverage that should be there is missing. Within the UK there is new legislation due on noise-at-work, within Europe there is new legislation on heavy metals in electrical systems and in the US, the Senate will shortly consider legislation that would require colleges and universities to provide fire safety information to prospective and current students.

Sometimes the shift in legislation is actually called for and supported by industry. In Australia, the New South Wales Government has recently passed the new Occupational Health and Safety Act 2000 and introduced new Occupational Health and Safety Regulations with effect from 1 September 2001. The new Act has replaced several existing laws (some nearly 100 years old), including the Occupational Health and Safety Act 1983 and regulations, the Construction Safety Act 1912 and regulations and parts of the Factories, Shops and Industries Act 1962 and regulations. The new Act and Regulations have been widely welcomed by industry groups, as it is seen that it updates and simplifies the laws relating to health and safety in the workplace. A significant advantage is that the new laws have been written in plainer language and contain new, specific provisions that require employers to have a greater understanding of their obligations to employees in relation to safety at work.

A safety case that still contains only the old requirements will be viewed as missing other details. It is a bad reflection on the safety case and the organisation responsible for it.

Mishap Occurrence in a Related Field

It need not be an incident with your product, inside your boundary diagram or at your site that leads to a review of your safety case (although, if this does happen, it should provoke a review anyway). If an incident happens in a related industry or with a similar product, it is very advisable to review the incident report (if you can get it) to see if your safety arrangements would have prevented the incident from happening.

Aside from this there are several other important reasons why you should read and learn from accident and incident reports [Holloway 2005];

- You will be less likely to believe the myths that are commonly believed concerning accident investigation and reporting.
- You will be more likely to have a realistic understanding of the potential consequences of 'extremely improbable' occurrences.
- You will have more courage to refuse to compromise safety if you encounter pressure to compromise.

Managing Change

Change is an inevitable part of the lifecycle of a system or product, and it should be planned for and managed in a systematic way. It is important that an adequate level of analysis is carried out to determine the safety impact of any change. To ensure that such changes are detected and addressed, a monitoring process should be implemented as part of the safety management system. A formal change control process should be put in place, such that a closed-loop system that provides feedback can be operated. Change information should be recorded (e.g. in the hazard log) to maintain configuration control over the system and to provide valuable data when managing future changes [MoD 2004].

Effective control is impossible without adequately documented instructions for how to actually do this, so it is important that a method for change management is developed and recorded. Accurate and timely information is critical to effective change management. All control systems are based upon the comparison of information about the situation as it is, with a predetermined standard of performance that has been called for in the original operation plan. Therefore, there is the requirement to have an on-going method for obtaining and communicating information about incidents and factors likely to affect risk and safety over the whole lifecycle of the system, process or product. A simple regular reporting format may be utilised noting any incidents or factors that will necessitate a change, and then a change-tracking matrix can be used to follow through the changes and provide the safety case (and managers) with the relevant change closed-loop feedback.

Since, as we have discussed, safety is important, perhaps the default assumption should be that change always impacts safety. Hence justification should be required for why it is not necessary to address safety as part of any change, rather than the other way around.

Review and Update Cycles

The maintenance of the safety case can only occur if there is a predetermined cycle of review and update that prompts the next look at the document and argument. Without this, even the best-intentioned safety team often 'forgets' or 'postpones' the review. Most often the incident that has been dreaded occurs during one of the 'postponed' time slots. The prompts for a safety case review should be listed out and agreed in the safety case document suite.

The normal recommendation for a review cycle is annually, which is fine for many safety cases, particularly where there has not been much changing in the operational environment and there haven't been any notable incidents. However, where change does take place, the safety case needs to be kept contemporary with the design and operational use of the product, process or system. Some industries are more stable than others, so the update cycle need not be so frequent. The particular frequency is best left to the safety engineers involved in each industry, as in some cases the time to review may even approach the 12 month! Additionally there are still the other mechanisms discussed above where a safety case review may be forced.

The program safety working group (PSWG) can be a specific forum for discussing change and monitoring operational factors. The PSWG can be given particular authority to ensure actions are carried out, and it should have a recording process in place to allow tracking of tasks and concerns. And although not necessarily attended by them, the group will have access to the senior managers of the organisation.

Notes

Holloway 2005, "Why you should read accident reports", NASA Langley Research Center, Software and Complex Electronic Hardware Standardization Conference, Norfolk Virginia, July 2005.

MoD 2004, "Safety Management Requirements for Defence Systems Part 1 Requirements", Interim Defence Standard 00:56 Issue 3, MoD, 2004.

Epilogue

Since this book was first thought about, and I started to write the chapters over the end of 2005 and into 2006, several more terrifying disasters have occurred in the world – e.g. the Buncefield explosion and fire, the Sago mining disaster and a number of military 'friendly-fire' incidents. They could have easily ended up as examples and discussion points within the pages, but at some point, you have to stop writing about current events and get the ideas down on the page before the next catastrophic event happens. Otherwise, the book would actually never get written – at the rate that accidents are (still!) happening, this book might not have been finished. Indeed some periodicals report on and discuss the accidents or convictions that have happened between publications, and they come out fortnightly!!

The safety domain may yet get stuck in and shake the corporate world awake – fortunately, some of it is already on the second coffee of the morning, having sorted through all the e-mails and letters. Regretfully, some of it is still curled up in bed, dreaming that the organisation is safe.

Thank you for reading, even buying this book, and still reading it at the very back, I cannot guarantee that any of us will not have an accident (even if you have read all the way to here), but if just one life is saved through someone doing something that they read between these covers, it will have been worthwhile.

Richard Maguire
September 2006
rlm@sevalidation.com

Index